X-LINKED TRAITS

X-Linked Traits
A Catalog of Loci in Nonhuman Mammals

JAMES R. MILLER

Biomedical Consultant, Central Research Division
Takeda Chemical Industries Ltd., Osaka, Japan

Honorary Professor, Department of Medical Genetics
University of British Columbia
Vancouver, B.C., Canada

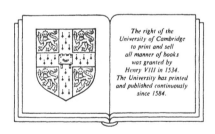

The right of the
University of Cambridge
to print and sell
all manner of books
was granted by
Henry VIII in 1534.
The University has printed
and published continuously
since 1584.

CAMBRIDGE UNIVERSITY PRESS

Cambridge
New York   Port Chester   Melbourne   Sydney

CAMBRIDGE
UNIVERSITY PRESS

32 Avenue of the Americas, New York NY 10013-2473, USA

Cambridge University Press is part of the University of Cambridge.

It furthers the University's mission by disseminating knowledge in the pursuit of education, learning and research at the highest international levels of excellence.

www.cambridge.org
Information on this title: www.cambridge.org/9780521373890

© Cambridge University Press 1990

First published 1990

A catalogue record for this publication is available from the British Library

Library of Congress Cataloguing in Publication data

Miller, James R., 1928–

X-linked traits.

Includes indexes.

1. Mammals – Genetic   2. Sex-linkage (Genetics)
3. X chromosome.   I. Title.   [DNLM: 1. Linkage
(Genetics).   2. X Chromosome – catalogs.   QH 600.5 M684x]
QL738.5.M55   1990   599´015   89–9937

ISBN 978-0-521-37389-0 Hardback

TO KIICHI TAKANO,
A GOOD FRIEND

CONTENTS

ACKNOWLEDGMENTS

The research work on this catalog was begun while I was on a study leave from the University of British Columbia in 1978–79. That leave was supported in part by a UBC Izaak Walton Killam Senior Memorial Fellowship, which I acknowledge with gratitude.

Subsequently, most of the research and writing was done while I was a biomedical consultant to the Central Research Division of Takeda Chemical Industries in Osaka. I thank Dr. K. Takano, to whom this book is dedicated, for arranging for this appointment, and Dr. E. Ohmura, Director of the Division when I joined Takeda, and his successors, Drs. K. Morita and Y. Sugino, for their support.

Among the many persons at Takeda who helped me, I mention four in particular: Dr. H. Iwasaki, a splendid office-mate in 1978–79, made my life easier than it might have been by introducing me to the way things are done in a Japanese company; he has continued to help me over the years. Mr. M. Yamazaki, who spent many hours patiently listening to my ideas on the catalog and then prepared a computer program for the text, never lived to see the fruits of his effort; he died suddenly on December 27, 1984. Dr. T. Murata, during a brief stint as head of the library at the Central Research Division, worked diligently to obtain some difficult references for me. Dr. Y. Kikuchi helped me with many matters during the early stages of preparing the text.

I also thank Dr. E. Matsunaga, director of the National Institute of Genetics, Mishima, for providing references.

A catalog such as this is based upon access to good libraries and the assistance of sympathetic, patient librarians. I was fortunate to have the generous support of the staff of the library of the Central Research Division throughout most of the year, and of Anna Leith and the marvelous staff of the Woodward Library of the University of British Columbia during each summer in Vancouver.

I express my appreciation to Drs. Diana Juriloff and Muriel Harris for constant support, many useful discussions, and valuable criticism of an early draft of the Introduction.

Sheila Manning, who helped Victor McKusick with the early editions of his catalog, ably assisted me until she retired and returned to England; she died on December 6, 1987. Over the years, Sachiko Noda and Alison Kelly provided excellent secretarial assistance which I appreciate greatly. Mary D. Miller helped me in many ways – translating, editing, typing, and providing advice – for which I am most grateful.

INTRODUCTION

BACKGROUND

Comparative studies have not played a central role in the development of mammalian genetics until quite recently. The reason for this is clear: There was little to compare with ease and certainty until spectacular advances in molecular biology revealed details of the genetic structure of many mammalian species. Coat color was the one major exception before the advent of molecular technologies. In the early decades of the century, much information accumulated on the genetics of coat colors in a wide array of mammalian species, and a few pioneers in genetics, including some of the outstanding figures in the history of the discipline, used this knowledge to speculate on the comparative aspects of mammalian genomes.

J. B. S. Haldane emerged as the dominant figure in this endeavor. In 1927, he published a famous paper in which he discussed the fundamentals of homology and how these could be demonstrated by examining the coat color genetics of six species of rodents and nine species of carnivores (Haldane 1927). His criteria for homology are presented, together with those of Lalley and McKusick (1985), in Table 1. Although the language and details understandably differ, the agreement on the fundamentals necessary for making a decision is impressive. Despite advances that have made the assignment of homology much more reliable, the uncertainty inherent in Haldane's criteria persists.

Haldane and others, notably Little (1958) and Searle (1961, 1968, 1969), recognized that the closer the phenotypes were to the primary action of the genes in question, the stronger any decision on homology would be. However, most of the phenotypes being studied and compared were far removed from primary gene effects and so could be influenced by many developmental processes, nongenetic as well as genetic. Despite recent dramatic advances in knowledge of the human and mouse genomes, which have led to the mapping of hundreds of loci – some with exquisite precision – in both species, the fact remains that many phenotypes being examined are gross clinical descriptions or, in the case of biochemical disorders, metabolic stages considerably removed from the primary effects of the responsible loci. Consequently, in addition to the firm homologous relations – based on detailed studies of nucleotide sequences in DNA or amino acid sequences in proteins – that accumulate apace, cruder phenotypes are continually being described whose possibly homologous relations must be considered even if definite proof of homology remains remote. Whereas true homologies are announced and examined with great en-

Table 1. *Comparison of the criteria as evidence for gene homology proposed by Haldane (1927) and Lalley and McKusick (1985)*

Haldane

Genes are homologous if they produce the same effects when brought in from either side in a species cross. (Experiments in vitro might yield equally definite results.) Homology may be suspected when any or all of the following criteria are satisfied:
1. The genes produce similar but not necessarily identical effects.
2. If a certain effect in species is produced by only 1 gene in each.
3. The genes have undergone several parallel mutations into more or less corresponding multiple alleles.
4. The genes exhibit similar linkages in different species.

Lalley and McKusick

The following criteria are recommended as evidence for identifying gene homologies:
1. Similar nucleotide or amino acid sequences.
2. Similar immunologic cross-reaction.
3. Formation of functional heteropolymeric molecules in interspecific somatic cell hybrids in cases of multimeric proteins.
4. Similar tissue distribution.
5. Similar developmental time of appearance.
6. Similar pleiotropic effects.
7. Similar substrate specificity.
8. Similar response to specific inhibitors.
9. Cross-hybridization to the same molecular probe.

Mapping or linkage homologies[a] in 2 or more species are considered well established if the following conditions are met:
1. Two or more pairs of homologous loci are linked or at least syntenic.
2. The order of 3 or more loci is the same.
3. The map distances among the loci are roughly equivalent.

[a]If the order is not the same, if another locus or segment is inserted in 1 species, or if the map distances are very different, then the term *syntenic homology* is preferred. Homologous linkage implies a conserved, undisrupted chromosome segment in 2 or more species because the species separated during evolution; syntenic homology does not imply conservation for the observed association of the loci.

thusiasm, the others tend to be neglected or disparaged, known or accepted by a few persons in a discipline such as veterinary neurology or by a larger group if the trait in question occurs in an animal such as the house mouse that has a larger body of adherents.

It has become extremely difficult to keep up with the genetics of the mouse and other mammals as knowledge accumulates rapidly. Publications such as the *Mouse News Letter* (Peters) and the *Rat News Letter* (Cramer) help one to keep up to date, and *Genetic Maps* (O'Brien 1987) makes comparing loci between species easier, but there is no central repository in which descriptions of phenotypes in one mammalian species and their possible relation to similar phenotypes in other mammalian species are stored. Mouse phenotypes are handily available in Margaret Green's splendid catalog (Green 1981), and the human being has been well served for many years by Victor McKusick's incomparable catalog, *Mendelian Inheritance in Man* (MIM), which is constantly updated and is now in its 7th edition (McKusick 1986). Comments on comparative aspects of phenotypes are scattered throughout MIM, and MIM5 has a secton on X chromosome homologies that does not appear in subsequent editions.

If comparative studies are to be encouraged and are not to be restricted to those loci that can be sequenced, mapped, or both, then there should be a place where the known genes (phenotypes) in a variety of nonhuman mammalian species are brought together and compared. The present volume is an attempt to treat mammalian X-linked traits in this manner.

Homology

The term *homology* derives from comparative anatomy, and its first use in this context is usually credited to Richard Owen, the distinguished English comparative anatomist, whose life spanned almost the nineteenth century. It is of interest that although the *Oxford English Dictionary* gives precedence to Owen for the use of *homologue* and *homologous*, it gives precedence to the English physician R. B. Todd for the use of *homology* in the first volume of his *The Cyclopaedia of Anatomy and Physiology,* published in June 1835. Owen himself (1847) stated that homolog was "first applied in anatomy by the German philosophical cultivators of that science." Probably the most elegant definition of homology is that given by the Medawars (1985): "Organs A, B, and C of genetically cognate structures are homologous if the development of one can be construed as a modification or variant of the development of another, or if the developments of all three are so many variants of the development of an evolutionary precursor of them all." The most succinct is given by van Valen (1982): "Homology is resemblance caused by a continuity of information."

In genetics, the term was first and is still most commonly used in relation to pairs of identical chromosomes called homologs. This use of homology, as well as its more specific and detailed use to describe genes and ultimately DNA sequences sharing a common origin, is understandable within the context of comparative anatomy if these molecular entities are considered as anatomic rather than biochemical units.

In the introduction to his paper, "The anatomy of the human genome," McKusick (1980: 370) states, "The gene map of the man's chromosome is part of his anatomy. Furthermore, there is a morbid anatomy, a comparative anatomy (with information relevant to evolution), a functional anatomy, a developmental anatomy, and even, if not a surgical anatomy, at least the beginnings of an applied anatomy." J. B. S. Haldane would not have been surprised by such developments. In his writings, he was awesomely prescient about the nature of the genic material, and his thinking and terminology anticipated current views by at least half a century. Haldane's (1927) argument was as follows: "Structures in two species are said to be homologous when they correspond to the same structure in a common ancestor." Because genes appear to be "definite structures in or on the chromosomes ... we are therefore justified in applying the principle of homology to them ... " (p. 199). Subsequently, he raised the question of whether a relation may exist between genes "more fundamental than homology, namely, chemical 'identity'." He pointed out that linkage studies suggested that genes were the same size as protein molecules. If this could be established, "comparative genetics and ultimately comparative morphology could be placed on a new basis" (p. 200).

Haldane could only speculate, but we can discuss the anatomy of the human genome and its evolutionary relations with those of other mammals with a confidence based on facts derived primarily from advances in molecular biology. However, the molecular approach has not resolved all issues. Colin Patterson of the British Museum is quoted as saying that "molecules aren't very different from morphology in this business, because they share similar problems" (Lewin 1985). But the fact remains that in comparative genetics the uncertainty becomes more acute the further removed the foci of comparison are from a gene and its primary product, and revolves around the same concern that confronted Haldane: What criteria are required to conclude with a comfortable degree of certainty that similar phenotypes in two or more mammalian species are true homologies? The difficulty in making a decision is most critical in the case of complex phenotypes stemming from pleiotropisms that result in syndromes. Searle (1968) was firm on this point; he wrote, "Pleiotropic effects tend to be variable and dependent on the genetic background, so no strict correspondence between phenotypes in different species can be expected even if the genes concerned are homologous" (p. 25).

Periodically debate flares up about the proper use of *homology* or *homologous*. Recently, a group of evolutionary biologists has deplored the fact that "homology" is used synonymously with "similarity" in many reports concerning sequence analysis of proteins and nucleic acids (Reeck et al. 1987). Such a debate is frequently sparked by the use of the term "homologous" to describe genes that produce the same or a strikingly familiar effect in two or more species without having any evidence that a common ancestor of the species possessed the gene. As Haldane (1927) pointed out, there can be no absolute criterion of genetic homology given the impossibility of studying ancestral forms; therefore, it seems reasonable and practical to use the term in the sense stated above, providing that the convention and its inadequacies are acknowledged. The same argument pertains to mutant alleles of apparently homologous genes – for example, Siamese dilution in the cat, and Himalayan dilution in the rabbit and mouse (Searle 1969: 28).

Gene nomenclature, gene symbols, and verbs

Nomenclature and the appropriate use of symbols remain formidable challenges to anyone interested in comparative mammalian genetics. Although an international committee has existed for many years to recommend nomenclature rules for mouse genetics (Dunn et al. 1940; Committee on Standardized Genetic Nomenclature for Mice 1980), and the *Mouse News Letter* (Peters) serves as a regular guide for appropriate usage, only a few such well-established, recognized, and respected committees and guides exist to set and maintain nomenclature standards in other species (Committee on Standardized Genetic Nomenclature for Cats 1968; Committee on Genetic Nomenclature of the Rat 1978). *The Council of Biology Editors (CBE) Style Manual* (1983) recommends that the rules proposed for the mouse "should be used for guinea pigs, hamsters, rabbits, rats, and white-footed mice *(Peromyscus)*." Some sense has been brought into human gene nomenclature as a consequence of rapid developments in gene mapping, which required that a semblance of order be established. However, as McKusick (1986) has pointed out, many nomenclature difficulties still persist, notably in relation to syndromes (MIM7: xxiii–xxv). Unfortunately, the system used in man and that used in the mouse and other mammals often differ in details, and the appeals of individuals who have devoted much thought to nomenclature issues go unheeded (Lyon 1987a). These, and other factors discussed below, make comparative nomenclature difficult and, at times, messy and confusing.

Adjectives such as "trivial," "fey," "whimsical," "capricious," "odd," and "arbitrary" are appropriate to describe the names often assigned to loci in experimental animals. Even when nomenclature guidelines exist, as they do for the mouse, some investigators ignore them and use their own systems; unfortunately, editors and reviewers do not challenge these idosyncratic or archaic deviations. (The number of instances in which "albino" is still used as the only description of mice in journals devoted to physiologic or toxicologic research is disturbing.) Investigators tend to use their own pet symbols for species for which no firm guidelines exist, or, more reasonably, attempt to follow a system used in another species, usually man or mouse. Given this disorder, it is not surprising that what appears to be a homologous mutation in two species is named differently in each (e.g., orange in the cat and tortoiseshell in the Syrian hamster – No. 31123 in this catalog); that the alpha-galactosidase locus (30150) is symbolized as *GLA, Gla, Aga,* or *Ags*; or that the same symbol is used to designate two quite different mutations in two species (e.g., *fa* for "fatty" in the rat and "falter" in the mouse).

Lalley and McKusick (1985) recommend that "as soon as the molecular basis of a phenotype is known in mouse, man and other species, the designation of the locus [should] be changed from the descriptive, often trivial, original designation to one that is molecularly based, so that homology can be evident from the symbol." They conclude, "It is very important that identical symbols be adopted in all species." This sound advice will require careful attention if it is to be implemented.

Although the nature of the material in this catalog makes it impossible to use a consistently uniform system of nomenclature, I have attempted to impose some order and have used either the human (McAlpine et al. 1987; Shows et al. 1987) or the mouse (Lyon 1985) system whenever possible.

On choosing a verb

Attempting to describe the relation between a gene and its related phenotype is fraught with hazards. Verbs such as "control" or "determine" may provide a reasonably accurate description of the rather direct relation between, say, a structural gene and an enzyme, but the greater the developmental or physiologic distance between a gene and its phenotype, the more inadequate these verbs seem. Gunther Stent, who has commented on the issue of the relation between gene and phenotype (Stent 1981, 1985), believes that "the difficulty in relating genome to phenome is not so much instrumental (in the sense that we still lack the observational means needed for incisive experiments) as it is conceptual (in the sense of capturing the essence of the phenomenon for which we are seeking an explanation)" (Stent 1984). Anyone who has studied the pathogenesis of a complex mammalian phenotype, such as that in a malformation syndrome, is well aware that, although a major gene effect may be involved, there is always evidence that its effect has been confounded by other genes (genetic background) or nongenetic factors. Under such circumstances, "initiate" may be a more appropriate verb than "determine" in the sense that the action of the major gene is the first essential step toward achieving the phenotype; however, its use may muddy rather than clarify an already murky area. Consequently, I have elected to retain "determine" in this catalog, while recognizing its inadequacies in conveying the vast complexities of the relation between gene and phenotype.

The mammalian X chromosome

Although Susumu Ohno is rightly associated with the concept of the conservation of the X chromosome during 150 million years of mammalian evolution (Ohno 1967, 1969), a distinguished animal geneticist, F. B. Hutt, first drew attention to the existence of apparently homologous X-linked mutations in humans and other mammals in 1953. The available evidence was limited to one locus in two species, but Hutt was astute enough to realize its potential significance. Later, Searle (1968: 250), who had access to more data, also commented on mammalian X-linked homologies.

Nevertheless, it was Ohno who not only critically assessed the significance of existing data but also formulated the hypothesis of the conservation of the mammalian X chromosome – the frozen X (Ohno 1973). The subsequent discovery of extensive autosomal conservation among mammalian species (Lalley and McKusick 1985) has not dislodged the X chromosome from its position as the model for understanding the conservation of linkage during the long period of mammalian evolution. The evidence on which Ohno based his law, though meager, was compelling. The prediction that the X chromosome of all mammalian species has the same genetic content has been sustained until quite recently when Cooper et al. (1984) and Dawson and Graves (1986) found that several cell hybrids retaining a marsupial X chromosome do not express marsupial steroid sulfatase (STS), and Sinclair et al. (1987) found that in two distantly related marsupials and a monotreme the ornithine transcarbamylase (OTC) locus is located on an autosome; the loci for both enzymes are X-linked (31125, 31341) in the few placental mammals studied to date. Sinclair et al. have proposed that an autosomal or pseudoautosomal segment containing OTC was translocated or incorporated into the inactivated region of the X chromosome quite recently in eutherian evolution, whereas the segment remained autosomal or was translocated to an autosome in metatherian or prototherian mammals.

Although much information has accumulated since Ohno first proposed his law, there is still a dearth of confirmed X-linked loci in a large number of different mammalian forms, and nothing resembling comparative gene maps has emerged save for those of mouse and man, which will be discussed further below. The numbers in Table 2 may convey the misleading impression that a large amount of information is available on X-linked loci in many mammalian species. If the data on the four loci (*Gla, G6pd, Hprt, Pgk*) presented in Table 4 are removed from Table 2, the number of nonhuman species is severely reduced and only two have more than five loci available to compare. Such numbers hardly provide a wealth of material on which to make profound statements on the comparative genetics of the mammalian X chromosome. As the summary in Table 2 indicates, the number of known X-linked loci in the mouse greatly exceeds that known in any other nonhuman mammal. Information on the location of the mouse loci appears in the individual entries in the body of the catalog. Linkage data on the mouse are maintained at the Jackson Laboratory in Bar Harbor, Maine by Davisson et al. (1988), and up-to-date maps are published regularly in the *Mouse News Letter* (Peters). Lyon (1988b) is compiling a mouse chromosome atlas that contains a genetic map of the mouse and the positions of known human homologs; it also shows the correlation between the genetic and cytogenetic maps. Nadeau and Reiner (1988) are compiling a genetic map of the mouse that shows the location of homologous loci mapped in both man and mouse. New techniques are being used to develop more detailed maps of the mouse X chromosome (Avner et al. 1987b; Brockdorff et al. 1987). Davisson (1987) published an excellent review of X-linked homologies in mouse and man. Linkage data on other mammalian species are available in *Genetic Maps* (O'Brien 1987), and an international committee on comparative mapping reports regularly to the Human Gene Mapping Workshops (Lalley et al. 1987).

Table 2. *Number of X-linked loci known (or suspected) in various mammals*

Human[a]	130 (164)
Mouse	66 (5)
Dog	12 (3)
Nonhuman primates (several species)	10 (1)
Cattle	9 (3)
Cat	9
Rat	9
Horse	7 (2)
Rabbit	6
Marsupials (several species)	6
Pig	5 (2)
Syrian hamster	5
Sheep	4
Chinese hamster	4
American mink	4
Indian muntjac	3
Red fox	3
Wood lemming	3

Notes: Only species (or groups) in which at least 3 loci are known are listed.

[a]Human data from McKusick VA: 1987. The human gene map. In: O'Brien SJ (ed), *Genetic Maps 1987: A Compilation of Linkage and Restriction Maps of Genetically Studied Organisms* (vol 4). Cold Spring Harbor, NY: Cold Spring Harbor Laboratory, p 535.

Although the genetic material of the X chromosome has been conserved, comparisons of the banding patterns of the X chromosome of a variety of mammalian species indicate that the chromosome has undergone extensive rearrangements during evolution, and an examination of the human and mouse gene maps supports this view. Buckle et al. (1985) examined the evolutionary relation of the X chromosomes of the two species; a summary of their conclusions, supplemented with data from Davisson (1987), Lyon (1988a), and other sources, is presented in Figure 1. A large conserved region, in which eight well-described homologous loci have been mapped, occupies the central portion of each chromosome. This region spans from Xp11–q11 to Xq21–q22 in the human and from D to F1 in the mouse. Outside these regions, there

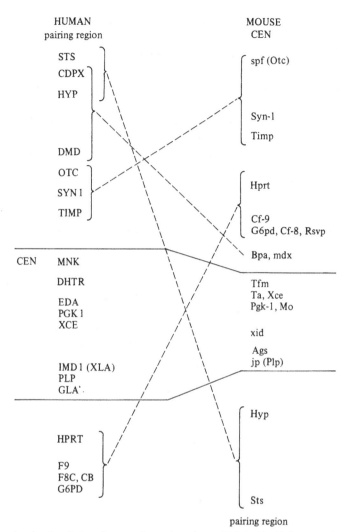

HUMAN
pairing region

STS
CDPX
HYP

DMD

OTC
SYN 1
TIMP

CEN   MNK

DHTR

EDA
PGK 1
XCE

IMD 1 (XLA)
PLP
GLA·.

HPRT

F9
F8C, CB
G6PD

MOUSE
CEN

spf (Otc)

Syn-1
Timp

Hprt

Cf-9
G6pd, Cf-8, Rsvp

Bpa, mdx

Tfm
Ta, Xce
Pgk-1, Mo

xid

Ags
jp (Plp)

Hyp

Sts

pairing region

Figure 1. Diagram showing the relation of apparently true homologous X-linked loci in human and mouse.
The positions of the loci within each chromosome are roughly relative; the principal intent is to illustrate
relations between the two chromosomes. An extensive evolutionarily conserved region exists in the central
portion of each chromosome. Outside of this region there are four other groups, each comprising two or
more loci.

Abbreviations: Ags, alpha-galactosidase; Bpa, bare patches; CB, several forms of color blindness; CDPX,
chondrodystrophia punctata; CEN, centromere; Cf-8, coagulation factor VIII; Cf-9, coagulation factor IX;
DHTR, dihydrotestosterone receptor; DMD, Duchenne muscular dystrophy; EDA, ectodermal dysplasia,
anhidrotic; F8C, coagulation factor VIII C; F9, coagulation factor IX; GLA, alpha-galactosidase; G6PD
(G6pd), glucose-6-phosphate dehydrogenase; HPRT (Hprt), hypoxantine phosphoribosyltransferase; HYP
(Hyp), hypophosphatemia, X-linked; IMD1, immunodeficiency 1, Bruton agammaglobulinemia; jp, jimpy;
mdx, muscular dystrophy, X-linked; MNK, the Menkes syndrome; Mo, mottled; OTC, ornithine
transcarbamylase; PGK1 (Pgk-1), phosphoglycerate kinase 1; Rsvp, red-sensitive visual pigment; spf,
sparse fur; SYN1 (Syn-1), synapsin I; Ta, tabby; Tfm, testicular feminization; TIMP (Timp), tissue inhibitor
of metalloproteinases; XCE (Xce), X-chromosome controlling element; xid, X-linked immune defect;
XLA, X-linked agammaglobulinemia. (Modified from Buckle et al. 1985; Davisson 1987; Lyon 1988a.)

are four other apparently homologous segments comprising two or more loci. In two instances the relation between the corresponding segments seems clear: a segment defined by *HPRT* (*Hprt*) and *G6PD* (*G6pd*), which lies at the extreme distal end of the long arm in the human and in A6–7 in the mouse, and a segment including *OTC*(*spf*), *SYN1* (*Syn-1*), and *TIMP* (*Timp*), which is located proximally in Xp in the human and distally in the mouse. The interspecies relations of the other three segments are more complex; clarification will depend on more detailed genetic analysis of the regions involved.

Davisson (1987) and Lyon (1988a) have demonstrated how as few as four rearrangements could explain the differences in the order of the loci on the two chromosomes. A minimum of two inversions – one moving the *HPRT–G6PD* segment and the other altering the relative position of the centromere – could explain the major differences. There now appear to be enough homologous markers spread out along the length of each chromosome to predict from the assignment of a newly described locus on one where its homolog might be expected on the other; examples of this are given by Davisson (1987) and Lyon (1988a) and also are found under specific entries in the catalog.

Avner et al. (1987a) used an interspecific mouse cross involving *Mus spretus* to show that five loci (*HPRT, FIX, DXS144, DXS52, G6PD*) in the Xq26–Xqter interval of the human X chromosome probably have the same order in the mouse X chromosome. Subsequently, Amar et al. (1987) used the same interspecific cross and cross-reacting molecular probes to localize 18 loci specific for the human and mouse X chromosomes. Their findings suggest that intrachromosomal rearrangements involving at least five X chromosome events must have occurred during the evolutionary separation of rodents and primates.

DNA fragments and segments

The number of cloned DNA sequences that have been isolated and used to identify specific chromosomal regions in man and the mouse is increasing dramatically. Fortunately, in both species, nomenclature rules have been adopted to designate these regions (Shows and McAlpine 1982: 221–222; Lyon 1985: 8). The human system uses alphanumeric symbols that are "locus-equivalent; each represents a unique chromosomal segment that can be identified by a cloned DNA sequence" (McAlpine et al. 1985: 9). Several X-linked fragments have been described in the mouse, and are being used intensively in mapping. The symbol for each one begins with "D" (for *D*NA), followed by "X" (for *X* chromosome), a symbol designating the laboratory that identified the marker (e.g., "Pas" for *Pas*teur Institute), and a number indicating the order of the marker mapped to the X chromosome in that laboratory (e.g., *DXPas1*). MIM does not list such segments described in man, and those in the mouse are not listed in this catalog proper but are described in Appendix I. In addition to those listed in Appendix I, Fisher et al. (1985) described ten random X-linked fragments, and

Avner and Bishop (1986) described three probes that identified sequences common to the X and Y chromosomes. The sequences in the latter are located in the extreme terminal region. Exchange of genetic material between the X and Y chromosomes has been demonstrated by use of all three probes. The first number of the *Mouse News Letter* each year lists DNA clones and probes.

General comments on the homologs in this catalog

Given that the available information varies considerably in quantity and quality from one entry to another, it is impossible to apply a set of uniform criteria for homology to all of them. I have tried to assess the evidence available for each entry and have accepted the opinions of those investigators who seem to have studied a given locus closely. In most cases, I believe the evidence supports their views. Consequently, there will probably be general agreement with most entries listed in Table 3. However, some readers may take issue with particular entries listed there and in Table 5, where the basis for decision making is uncertain and perhaps controversial.

Tables 3, 5, and 6 list, respectively, 32 human loci for which homologies are considered proved to exist in at least one other mammalian species (for each of *GLA, G6PD, HPRT,* and *PGK* more than 25 species are known, Table 4); a minimum of an additional 13 loci for which probable or possible homologies exist in nonhuman mammals; and 44 loci known to be X-linked in species other than man that have no obvious human homologs at present. The 12 entry headings listed in Table 7 are termed "exceptional" because I cannot fit them easily into other categories; the reasons they are classified this way are made clear in the respective entries.

The 44 entries listed in Table 6 merit comment. One (30188) is a tail disorder and seven (30413, 30645, 30898, 31123, 31156, 31353, 31499) concern loci that determine coat condition or coat color; six of the latter are in the mouse, and one is in the Syrian hamster and the cat. To my knowledge, the molecular basis of none of these eight loci is known, but the fine details of the X chromosome of man (Davies 1985; Drayna and White 1985) and mouse (Fisher et al. 1985) are being analyzed rapidly, and it is possible that some form of molecular sequencing involving certain of the six mouse loci will emerge soon. Such analyses will make it possible to look for homologous sequences in the human genome and to assess their function in man.

Several other entries in Table 6 include hematologic (30154), metabolic (30432), structural (30196, 30337, 31202), or other disorders (30214, 30549, 31433) for which human homologs, if they exist, should be easily detected.

Seven other loci in Table 6 (30694, 30821, 30822, 30826, 30827, 30906, 31434) controlling immune responses may be homologous with the several known X-linked loci that appear to control immunologic functions in man.

Finally, three entries concern loci involved with host reactions to drugs (30336) or viruses (30838, 31236); these are particularly difficult to investigate in man (Zabriskie and Gibofsky 1986).

Table 3. *Apparently true X-linked homologous loci or traits (32) in man and other mammals*

MIM No.	MIM name	Species
30030	AGAMMAGLOBULINEMIA	cow?, horse, mouse
30150[a]	ANGIOKERATOMA, DIFFUSE (GLA DEFICIENCY)	>25 species
30179[b]	ARAF PROTOONCOGENE	mouse
30295	CHONDRODYSPLASIA PUNCTATA, X-LINKED	mouse
30380[c]	COLOR BLINDNESS, PARTIAL, DEUTAN SERIES	squirrel monkey, mouse
30390[c]	COLOR BLINDNESS, PARTIAL, PROTAN SERIES	
30510	ECTODERMAL DYSPLASIA, ANHIDROTIC	cow, dog?, mouse
30590	GLUCOSE-6-PHOSPHATE DEHYDROGENASE (G6PD)	>30 species
30600[d]	GLYCOGEN STORAGE DISEASE VIII (HEPATIC PHOSPHORYLASE KINASE DEFICIENCY)	mouse
30640[e]	GRANULOMATOUS DISEASE, CHRONIC	mouse
30670[f]	HEMOPHILIA A	cat, cow?, dog, horse
30690[g]	HEMOPHILIA B	cat, dog, mouse
30697	H–Y REGULATOR	wood lemming
30780	HYPOPHOSPHATEMIA, X-LINKED	mouse
30781	HYPOPHOSPHATEMIA B	mouse
30800	HYPOXANTHINE PHOSPHORIBOSYL-TRANSFERASE (HPRT)	>25 species
30810[h]	ICHTHYOSIS, X-LINKED (STEROID SULFATASE DEFICIENCY)	mouse, wood lemming
30820	ICHTHYOSIS AND MALE HYPOGONADISM	mouse
30940	MENKES SYNDROME	hamster, mouse
31010	MUSCULAR DYSTROPHY, DUCHENNE AND BECKER TYPES (DMD)	dog, mouse
31099[i]	ONCOGENE HARVEY RAS-2	cat, rat
31125	ORNITHINE CARBAMOYLTRANSFERASE	mouse, rat
31135	OUBAIN RESISTANCE	mouse

Table 3. (*cont.*)

MIM No.	MIM name	Species
31180	PHOSPHOGLYCERATE KINASE (PGK)	>30 species
31208[j]	PROTEOLIPID PROTEIN, MYELIN	dog, mouse, rat
31344	SYNAPSIN 1	mouse
31365	TEMPERATURE-SENSITIVE MUTATION	hamster, mouse
31370[k]	TESTICULAR FEMINIZATION SYNDROME (ANDROGEN RECEPTOR DEFICIENCY)	chimpanzee, cow, horse, mouse, pig?, rat
31386[l]	TESTIS-DETERMINING FACTOR-LIKE SEQUENCES	cow, dog, goat, gorilla, horse, owl, rabbit, rhesus monkey
31420	THYROXINE-BINDING GLOBULIN OF SERUM	baboon
31422[m]	TISSUE INHIBITOR OF METALLOPROTEINASES	mouse
31467	X-CHROMOSOME CONTROLLING ELEMENT	kangaroo, mouse
31470	XG BLOOD GROUP SYSTEM	gibbon

[a]ALPHA GALACTOKINASE (GLA) in this catalog.

[b]Number and name are for this catalog, as MIM number and name are not assigned. McAlpine et al. (1987) state 31101.

[c]1453, VISUAL PIGMENTS, X-LINKED in this catalog.

[d]31187, PHOSPHORYLASE KINASE in this catalog.

[e]30433, CYTOCHROME b-245 BETA POLYPEPTIDE in this catalog.

[f]30356, COAGULATION FACTOR VIII in this catalog.

[g]30358, COAGULATION FACTOR IX in this catalog.

[h]31341, STEROID SULFATASE in this catalog.

[i]30654, HARVEY SARCOMA PROTOONCOGENE 2 in this catalog.

[j]This is the same locus as 31160, Pelizaeus–Merzbacher disease.

[k]30494, DIHYDROTESTOSTERONE RECEPTOR in this catalog.

[l]Number and name are for this catalog as MIM number and name are not assigned.

[m]Number and name are for this catalog, as MIM number and name are not assigned. McAlpine et al. (1987) state 30537, ERYTHROID POTENTIATING ACTIVITY.

Source: McAlpine PJ, Van Cong N, Boucheix C, Pakstis AJ, Doute RC, Shows TB: 1987. The 1987 catalog of mapped genes and report of the nomenclature committee. *Cytogenet Cell Genet* 46: 29–101.

Table 4. *Species in which GLA, G6PD, HPRT, or PGK is known to be X-linked*

	GLA	G6PD	HPRT	PGK
African green monkey	+	+	−	+
American mink	+	+	+	+
Cat	+	+	+	−
Chimpanzee	+	+	−	+
Chinese hamster	−	+	+	+
Cattle	+	+	+	+
Dasyurids	+	+	+	+
Dog	+	+	+	−
Gibbon	+	−	+	+
Gorilla	+	+	−	+
Hare	−	+	−	−
Horse	−	+	+	+
Indian muntjac	−	+	+	+
Macropodids	+	+	+	+
Mouse	+	+	+	+
Mouse lemur	+	+	+	+
Orangutan	+	+	−	−
Owl monkey	+	+	−	+
Pig	+	+	+	+
Rabbit	+	+	+	+
Rat	+	+	+	+
Red fox	+	+	+	−
Rhesus monkey	+	+	−	−
Sheep	+	+	+	+
Virginia opossum	−	+	−	+
Vole	−	+	+	−

The category "loci known in man but not in other mammals" is extensive and includes a variety of genes that give rise to phenotypes difficult to define and detect in most mammalian species; mental retardation is the most obvious of these. Of the 294 loci (McKusick 1987) known (or suspected) to be X-linked in the human, 11 (4%) involve mental retardation primarily.

McKusick (1980: 384) points out that one of the most obvious reasons for the difference in the number of X-linked traits in man and mouse is the large number of "paralogous" loci that may have arisen by tandem duplication in the former. [*Paralogous homology* refers to the situation in which two or more independent genes

Table 5. *Probable or possible X-linked homologies (15, representing a minimum of 13 loci)*

Catalog		Animal	MIM7 Number	Name
Number	Name			
30221	CATARACT, X-LINKED	mouse	30220	CATARACT, CONGENITAL TOTAL, WITH POSTERIOR SUTURAL OPACITIES IN HETEROZYGOTES
30237	CELL SURFACE ANTIGEN, X-LINKED	red kangaroo	31345, 31346, 31347	VARIOUS CELL SURFACE ANTIGENS
30403	CONGENITAL TREMOR TYPE IIIA	pig	31160	PELIZAEUS–MERZBACHER DISEASE
30536	EYE–EAR REDUCTION	mouse	30970	MICROPHTHALMIA
30561	FRAGILE SITE XC–D	mouse	a^2	FRAGILE SITE Xq22
30655	HEMOGLOBIN PRODUCTION, INHIBITION OF	mouse	30130	ANEMIA, HYPOCHROMIC
30692	HINDLEG PARALYSIS	Syrian hamster	30280	CHARCOT–MARIE–TOOTH PERONEAL MUSCULAR ATROPHY

Table 5 (*cont.*)

	Catalog			MIM7	
Number	Name		Animal	Number	Name
30856	IRREGULAR TEETH		mouse	31350	TEETH, ABSENCE OF
31048	NEPHRITIS, HEREDITARY, X-LINKED		dog	30105	ALPORT SYNDROME
31147	PARALYTIC TREMOR		rabbit	31150	PARKINSONISM
31343	STREAKED HAIRLESSNESS		cow	30560	FOCAL DERMAL HYPOPLASIA or INCONTINENTIA PIGMENTI
31347	SURFACE ANTIGEN MIC2		chimpanzee and gorilla[b]	31347	SURFACE ANTIGEN MIC2
31351	STRIATED		mouse	30830	FOCAL DERMAL HYPOPLASIA or INCONTINENTIA PIGMENTI
31352	TATTERED		mouse	30830	INCONTINENTIA PIGMENTI
31491	XY SEX REVERSAL		horse	30610	GONADAL DYSGENESIS, XY, FEMALE TYPE (SWYER SYNDROME)

[a]Not listed in MIM.
[b]X-linkage not proven in these species.

Table 6. *X-linked loci (44) described in nonhuman mammals but not known in man*

Catalog		
Number	Name	Species
30154	ANEMIA, X-LINKED	mouse
30184	BALB VIRUS RESTRICTION	cat
30188	BENT TAIL	mouse
30196	BROAD HEADED	mouse
30214	CARPAL SUBLUXATION	dog
30234	cDNA pMIF 3/10 INDUCED WITH INTERFERON	mouse
30236	CELL SURFACE ANTIGEN VP382	cat
30293	CHEMOSENSORY IDENTITY	mouse
30336	CLEFT PALATE, CORTISONE-INDUCED, LIABILITY TO	mouse
30337	CLEFT PALATE, POLYDACTYLY, SYNDACTYLY, TIBIOFIBULAR SHORTENING, BRACHYGNATHISM, AND SCOLIOSIS	dog
30413	CREAM	mouse
30432	CYSTINURIA	dog
30549	FIDGET, X-LINKED	mouse
30554	FRAGILE SITE	cattle
30556	FRAGILE SITE fra(X)(q1), FOLATE-SENSITIVE	Indian mole rat
30557	FRAGILE SITE fra(X)(q2), FOLATE-SENSITIVE	Indian mole rat
30558	FRAGILE SITE fra(X)(q3), FOLATE SENSITIVE	Indian mole rat
30559	FRAGILE SITE fra(X)(cen), FOLATE-SENSITIVE	Indian mole rat
30645	GREASY	mouse
30653	HARLEQUIN	mouse
30694	HISTOCOMPATIBILITY-X	mouse, rat
30821	IMMUNE RESPONSE TO DENATURED DNA	mouse
30822	IMMUNE RESPONSE TO LDH-C_4, TEMPORAL REGULATION OF	mouse
30826	IMMUNE RESPONSE TO SYNTHETIC DOUBLE-STRANDED RNA	mouse

Table 6 (*cont.*)

Catalog		
Number	Name	Species
30827	IMMUNE RESPONSE TO TYPE III PNEUMOCOCCAL POLYSACCHARIDE	mouse
30838	INTERFERON (IFN) PRODUCTION, EARLY, VIRUS-INDUCED	mouse
30898	LINED	mouse
30906	LYMPHOCYTE-REGULATED, X-LINKED	mouse
30986	MOLONEY MURINE LEUKEMIA VIRUS INTEGRATION SITE-14	mouse
31047[a]	NADH–COENZYME Q REDUCTASE DEFICIENCY (ELECTRON TRANSPORT CHAIN, DEFECT OF COMPLEX I OF)	Chinese hamster, mouse
31123	ORANGE, TORTOISESHELL[b]	cat, Syrian hamster
31124	ORNITHINE DECARBOXYLASE, X-LINKED	mouse
31156	PATCHY COAT	mouse
31202	POLYDACTYLY, PREAXIAL, WITH HEMIMELIA AND UROGENITAL DEFECTS, X-LINKED	mouse
31236	RAUSCHER LEUKEMIA VIRUS SUSCEPTIBILITY-3	mouse
31276	RIBONUCLEIC ACID, 7S	mouse
31288	SEGREGATION REVERSAL	mouse
31353	TAWNY	Syrian hamster
31354	TEAR PROTEIN SYSTEM-3	mouse
31433	TREMBLY	mouse
31434	TRF-ACCEPTOR SITE(S) ON B LYMPHOCYTES, EXPRESSION OF	mouse
31493	X–Y UNIVALENCY, X-LINKED FACTORS CONTROLLING	mouse
31498	Y CHROMOSOMAL SEQUENCES	mouse
31499	YELLOW MOTTLING	mouse

[a]Although listed in MIM7 (31245), the existence of this locus in man is not proved.
[b]Homologs.

Table 7. *Exceptional entries not easily classified*

	Catalog	
Number	Name	Species
30006	ABSENT PINNAE	black rhinoceros
30070	ALBINISM–DEAFNESS SYNDROME	dog
30615	GONAD-SPECIFIC RECEPTOR OF H–Y ANTIGEN	wood lemming
30702	HYDROCEPHALUS, X-LINKED	rat
30825	IMMUNOGLOBULIN M, LEVEL OF	mouse
30892	LETHAL, X-LINKED, 1	mouse
30894	LETHAL X-LINKED, 2	mouse
30987	MOLONEY MURINE LEUKEMIA VIRUS INTEGRATION SITE	mouse
31201	POLYDACTYLY	cattle
31217	QRS INTERVAL	horse
31336	SPLAYLEG	pig
31384	TESTIS ASCORBIC ACID	mouse

in an organism have arisen from a common ancestral gene by duplication or translocation and subsequent divergence (Fitch 1976: 161).] In addition to the large number of loci involved with mental retardation, the human X chromosome carries three loci for ocular albinism, five for end-organ responsiveness, and at least four for immunodeficiencies; several forms of retinitis pigmentosa and retinal dysplasia, and several spinocerebellar degenerations and possibly related central nervous system disorders are X-linked; PRPP synthetase and HPRT, enzymes for purine metabolism, are X-linked; and *Xk*, the structural gene for Kell blood group precursor substance is closely linked to *Xg*. In the mouse, only a few such clusters have been detected: three loci for end-organ responsiveness (two forms of hereditary hypophosphatemia, and the dihydrotestosterone receptor); two loci affecting absorption of mineral elements (*sla*, a locus affecting intestinal absorption of iron, and *Mo*, a locus affecting copper absorption); several loci concerned with immune responsiveness; and two closely linked loci, one of which is definitely, the other possibly, concerned with steroid metabolism.

The current status of some of the homologs cataloged here is in doubt. Whereas homologs in rodents and other small laboratory mammals are usually treasured and can be maintained quite easily, those in medium-sized or large domestic mammals create difficulties: The affected animals are usually economically worthless and are expensive to maintain unless an exceptional effort is made to provide adequate facilities; many models are undoubtedly lost through funding cutbacks (Barnes 1986). Within recent years, even a large well-established center "restructured" its rabbit genetics program by discontinuing research on rabbits and scaling down rabbit genetic resources; the only redeeming aspect of the exercise was that stock for inbred lines and valuable mutants are now maintained as either frozen embryos or sperm. Cryobiology may provide the means by which valuable homologs are maintained until they can be studied properly (Yoshiki et al. 1987), but the techniques involved can be expensive if large numbers of specimens are to be stored over long periods. Unfortunately, these techniques are not widely available, and undoubtedly some of the homologs included in this catalog are extinct. I have not tried strenuously to confirm this point and eliminate those that are now extinct, because I believe some readers may find the available information useful, and, in addition, I hope knowledge of the existence and potential usefulness of such traits will contribute in some small way to an awareness of the need to preserve such valuable biologic material.

In the foreword to *The Second Tree from the Corner*, a collection of his essays, E. B. White wrote, "I think there has never been an age more cruel to writers than this one – rendering their stuff obsolete almost before it escapes from the typewriter." White's concerns are trivial when compared to the perils of instantaneous obsolescence faced by anyone attempting to organize a synthesis of the type presented here. In addition to this hazard, I recognize that there will be errors of omission and commission. I encourage anyone who uses this catalog to draw my attention not only to frank errors but to inappropriate interpretations of results and conclusions relating to the homologies described.

September 30, 1988
James R. Miller
Osaka, Japan

References

Amar L, Dandalo L, Arnaud D, Hanauer A, Ryder-Cook A, Mandel JL, Avner P: 1987. Conservation and reorganisation of loci on the mammalian X chromosome: a molecular framework for the identification of homologous sub-chromosomal regions in man and mouse. (Abstr) *Mouse News Lett* 79: 56 only.

Avner PR, Bishop C: 1986. Private communication. *Mouse News Lett* 74: 98 only.

Avner P, Amar L, Arnaud D, Hanauer A, Cambrou J: 1987a. Detailed ordering of markers localizing to the Xq26–Xqter region of the human X chromosome by the use of an interspecific *Mus spretus* mouse cross. *Proc Natl Acad Sci USA* 84: 1629–1633.

Avner P, Arnaud D, Amar L, Cambrou J, Winking H, Russell LB: 1987b. Characterization of a panel of somatic cell hybrids for regional mapping of the mouse X chromosome. *Proc Natl Acad Sci USA* 84: 5330–5334.

Barnes DM: 1986. Tight money squeezes out animal models. *Science* 232: 309–311.

Brockdorff N, Fisher EMC, Cavanna JS, Lyon MF, Brown SDM: 1987. Construction of a detailed molecular map of the mouse X chromosome by microcloning and interspecific crosses. *EMBO J* 6: 3291–3297.

Buckle VJ, Edwards JH, Evans EP, Jonasson JA, Lyon MF, Peters J, Searle AG: 1985. Comparative maps of human and mouse X chromosomes. (Abstr) *Cytogenet Cell Genet* 40: 594–595.

CBE Style Manual Committee: 1983. *CBE Style Manual* 5th ed. Bethesda, MD: Council of Biology Editors, p 193.

Committee on Standardized Genetic Nomenclature for Cats: 1968. Standardized genetic nomenclature for the domestic cat. *J Hered* 59: 39–40.

Committee on Standardized Genetic Nomenclature for Mice: 1980. New rules for nomenclature of genes, chromosome anomalies and inbred strains. *Mouse News Lett* 61: 4–16. (Rules for gene nomenclature reprinted in: Green MC (ed): 1981. *Genetic Variants and Strains of Laboratory Mouse.* Stuttgart: Gustav Fischer Verlag, pp 1–7.)

Committee on Genetic Nomenclature of the Rat: 1978. *Rat News Lett* 4: 10–11.

Cooper DW, McAllen BM, Donald JA, Dawson G, Dobrovic A, Graves JAM: 1984. Steroid sulphatase is not detected on the X chromosome of Australian marsupials. (Abstr) *Cytogenet Cell Genet* 37: 439 only.

Cramer DV (ed): *Rat News Letter* (published twice a year in January and July).

Davies KE: 1985. Molecular genetics of the human X chromosome. *J Med Genet* 22: 243–249.

Davisson MT: 1987. X-linked genetic homologies between mouse and man. *Genomics* 1: 213–227.

Davisson MT, Roderick TH, Hillyard AL, Doolittle DP: 1988. Linkage map of the mouse. *Mouse News Lett* 81: 12–19.

Dawson GW, Graves JAM: 1986. Gene mapping in marsupials and monotremes. III. Assignment of four genes to the X chromosome of the wallaroo and the euro (*Macropus robustus*). *Cytogenet Cell Genet* 42: 80–84.

Drayna D, White R: 1985. The genetic linkage map of the human X chromosome. *Science* 230: 753–758.

Dunn LC, Gruneberg H, Snell GD: 1940. Report of the Committee on Mouse Genetics Nomenclature. *J Hered* 31: 505–506.

Fisher EMC, Cavanna JS, Brown SDM: 1985. Microdissection and microcloning of the mouse X chromosome. *Proc Natl Acad Sci USA* 82: 5846–5849.

Fitch WM: 1976. Molecular evolutionary clocks. In: Ayala FJ (ed), *Molecular Evolution.* Sunderland, MA: Sinauer Associates, pp 160–178.

Green MC: 1981. Catalog of mutant genes and polymorphic loci. In: Green MC (ed), *Genetic Variants and Strains of Laboratory Mice.* Stuttgart: Gustav Fischer Verlag, pp 8–278.

Haldane JBS: 1927. The comparative genetics of colour in Rodents and Carnivora. *Biol Rev* 2: 199–212.

Hutt FB: 1953. Homologous sex-linked mutations in man and other mammals. *Am Nat* 87: 160–162.

Lalley PA, McKusick VA: 1985. Report of the Committee on Comparative Mapping. *Cytogenet Cell Genet* 40: 536–566.

Lalley PA, O'Brien SJ, Creau-Goldberg N, Davisson MT, Roderick TH, Echard G, Womack JE, Graves JM, Doolittle DP, Guidi JN: 1987. Report of the Committee on Comparative Mapping. *Cytogenet Cell Genet* 46: 367–389.

Lewin R: 1985. Molecules vs. morphology: of mice and men. *Science* 229: 743–745.

Little CC: 1958. Coat color genes in rodents and carnivores. *Q Rev Biol* 33: 103–137.

Lyon MF: 1985. Rules for nomenclatures of genes, chromosome anomalies and inbred strains. *Mouse News Lett* 72:2–20.

Lyon MF: 1987a. Nomenclature for homeo-box-containing genes. *Nature* 325: 22 only.

Lyon MF: 1988a. The Willam Allan Memorial Award Address: X-chromosome inactivation and the location and expression of X-linked genes. *Am J Hum Genet* 42: 8–16.

Lyon MF: 1988b. Mouse chromosome atlas. *Mouse News Lett* 81:20–41.

McAlpine PJ, Shows TB, Miller RL, Pakstis AJ: 1985. The 1985 catalog of mapped genes and report of the nomenclature committee. *Cytogenet Cell Genet* 40: 8–66.

McAlpine PJ, Van Cong N, Boucheix C, Pakstis AJ, Doute RC, Shows TB: 1987. The 1987 catalog of mapped genes and report of the nomenclature committee. *Cytogenet Cell Genet* 46: 29–101.

McKusick VA: 1978. *Mendelian Inheritance in Man: Catalogs of Autosomal Dominant, Autosomal Recessive, and X-Linked Phenotypes*, 5th ed. Baltimore: Johns Hopkins Univ Press, pp lxxxvii–xci.

McKusick VA: 1980. The anatomy of the human genome. *J Hered* 71: 370–391.

McKusick VA: 1986. *Mendelian Inheritance in Man: Catalogs of Autosomal Dominant, Autosomal Recessive, and X-linked Phenotypes*, 7th ed. Baltimore: Johns Hopkins Univ Press.

McKusick VA: 1987. The human gene map. In: O'Brien SJ (ed), *Genetic Maps 1987: A Compilation of Linkage and Restriction Maps of Genetically Studied Organisms* (vol 4). Cold Spring Harbor: Cold Spring Harbor Laboratory, p 535.

Medawar P, Medawar J: 1985. *Aristotle to Zoos: A Philosophical Dictionary of Biology.* Oxford: Oxford Univ Press, p 145.

Nadeau JH, Reiner A: 1988. Map of mouse and man homologies. *Mouse News Lett* 81: 42–44.

O'Brien SJ (ed): 1987. *Genetic Maps 1987: A Compilation of Linkage and Restriction Maps of Genetically Studied Organisms* (vol 4). Cold Spring Harbor: Cold Spring Harbor Laboratory, pp 430–517.

Ohno S: 1967. *Sex Chromosomes and Sex Linked Genes*. Berlin: Springer-Verlag, pp 46–73.

Ohno S: 1969. Evolution of sex chromosomes in mammals. *Annu Rev Genet* 3: 495–524 (pp 496–501).

Ohno S: 1973. Ancient linkage groups and frozen accidents. *Nature* 244: 259–262.

Owen R: 1847. Report on the archetype and homologies of the vertebrate skeleton. *Br Assoc Adv Sci Rep* 16 (for 1846): 169–340 (p 173).

Peters J (ed): *Mouse News Letter*. Oxford: Oxford Univ Press. (Published three times a year in February, July, and November; the third issue is devoted primarily to a listing of inbred strains.)

Reeck GR, de Haen C, Teller DC, Doolittle RF, Fitch WM, Dickerson RE, Chambon P, McLachlan AD, Margoliash E, Jukes TH, Zuckerkandl E: 1987. "Homology" in proteins and nucleic acids: a terminology muddle and a way out of it. *Cell* 50: 667 only.

Searle AG: 1961. Comparative genetics. In: Gedda L (ed), *De Genetica Medica* (vol 1). Rome: Instituto Mendel, pp 185–241.

Searle AG: 1968. *Comparative Genetics of Coat Colour in Mammals*. London: Logos Press.

Searle AG: 1969. Coat color genetics and problems of homology. In: Dronamraju KR (ed), *Haldane and Modern Biology*. Baltimore MD: Johns Hopkins Univ Press, pp 27–41.

Shows TB, McAlpine PJ: 1982. The 1981 catalogue of assigned human genetic markers and report of the nomenclature committee. *Cytogenet Cell Genet* 32: 221–245.

Shows TB, McAlpine PJ, Boucheix C, Collins, FS, Conneally PM, Frezal J, Gershowitz H, Goodfellow PN, Hall JG, Issitt P, Jones CA, Knowles BB, Lewis M, McKusick VA, Meisler M, Morton NE, Rubinstein P, Schanfield MS, Schmickel RD, Skolnick MH, Spence MA, Sutherland, GR, Traver M, Van Cong N, Willard HF: 1987. Guidelines for human gene nomenclature: an international system for human gene nomenclature (ISGN 1987). *Cytogenet Cell Genet* 46: 11–28.

Sinclair AH, Wrigley JM, Graves JAM: 1987. Autosomal assignment of OTC in marsupials and monotremes: implications for the evolution of sex chromosomes. *Genet Res* 50: 131–136.

Stent GS: 1981. Strength and weakness of the genetic approach to the development of the nervous system. *Annu Rev Neurosci* 4: 163–194.

Stent GS: 1984. From probability to molecular biology. (A review of Slack JMW: *From Egg to Embryo*) *Cell* 36: 567–570.

Stent GS: 1985. Thinking in one dimension: the impact of molecular biology on development. *Cell* 40: 1–2.

van Valen LM: 1982. Homology and causes. *J Morphol* 173: 305–312.

Yoshiki A, Ohno K, Wakasugi N: 1987. Cryopreservation of strains and mutant genes in mice. *Exp Anim* 36: 379–386.

Zabriskie JB, Gibofsky A: 1986. Genetic control of the susceptibility to infection with pathogenic bacteria. *Curr Top Microbiol Immunol* 124: 1–20.

APPENDIX I:
X-LINKED DNA SEGMENTS IN THE MOUSE

Appendix I: *X-linked DNA segments in the mouse*

Locus	Location	Human homolog	Reference
DXPas1	midway between *Plp* and *Xpl*	—	Amar et al. 1985
DXPas2	between *Pgk-1* and *Aga*	—	Amar et al. 1985
DXPas3	proximal to *DXPas4*	—	Amar et al. 1985
DXPas4	proximal to *DXPas5*	—	Amar et al. 1985
DXPas5	proximal to *Hprt*	—	Amar et al. 1985
DXPas6	between *Cf-9* and *G6pd*	*DXS144*	Avner et al. 1987a
DXPas7	between *spf* and *Hprt*	—	Avner et al. 1987b
DXPas8	between *G6pd* and *Ta*	*DXS52*	Avner et al. 1987a
DXPas9	—	—	Unpublished;
DXPas10	—		listed in *Mouse News Lett* (1987)
DXPas11	—	—	77: 88 only
DXPas12	—	—	
DX17A	—	—	Disteche et al. 1982
DX31A	—	—	Disteche et al. 1982
DX6836	near centromere	—	Disteche et al. 1985
DX7038	near centromere	—	Disteche et al. 1985

Note: In addition to the loci listed here, Mandel et al. (1986) mention 6 anonymous X-linked sequences detected in both man and mouse, and a human locus, *DXS32,* which has an unnamed mouse homolog (Table 2). The mouse homolog maps close to *spf* (Avner 1986), and *DXS32* is closely linked to *OTC* (the human homolog to *spf*) (de Martinville et al. 1985).

References

Amar LC, Arnaud D, Cambrou J, Guenet J-L, Avner PR: 1985. Mapping of the mouse X chromosome using random genomic probes and an interspecific mouse cross. *EMBO J* 4: 3695–3700.

Avner PR: 1986. Private communication. *Mouse News Lett* 74: 98 only.

Avner P, Amar L, Arnaud D, Hanauer A, Cambrou J: 1987a. Detailed ordering of markers localizing to the Xq26–Xqter region of the human X chromosome by the use of an interspecific *Mus spretus* mouse cross. *Proc Natl Acad Sci USA* 84: 1629–1633.

Avner P, Arnaud D, Amar L, Cambrou J, Winking H, Russell LB: 1987b. Characterization of a panel of somatic cell hybrids for regional mapping of the mouse X chromosome. *Proc Natl Acad Sci USA* 84: 5330–5334.

de Martinville B, Kunkel LM, Bruns G, Morle F, Koenig M, Mandel JL, Horwich A, Latt SA, Gusella JF, Housman D, Francke U: 1985. Localization of DNA sequences in region of Xp21 of the human X chromosome: search for molecular markers close to the Duchenne muscular dystrophy locus. *Am J Hum Genet* 37: 235–249.

Disteche CM, Kunkel LM, Lojewski A, Orkin SH, Eisenhard M, Sahar E, Travis B, Latt SA: 1982. Isolation of mouse X-chromosome specific DNA from an X-enriched lambda phage library derived from flow-sorted chromosomes. *Cytometry* 2: 282–286.

Disteche CM, Tantravahi U, Gandy S, Eisenhard M, Adler D, Kunkel LM: 1985. Isolation and characterization of two repetitive DNA fragments located near the centromere of the mouse X chromosome. *Cytogenet Cell Genet* 39: 262–268.

Mandel JL, Arveiler B, Camerino G, Hanauer A, Heilig R, Koenig M, Oberle I: 1986. Genetic mapping of the human X chromosome: linkage analysis of the q26–q28 region that includes the fragile X locus and isolation of expressed sequences. *Cold Spring Harbor Symp Quant Biol* 51: 195–203.

APPENDIX II:
THE MAMMALIAN Y CHROMOSOME

The mammalian Y chromosome, which had received little detailed attention previously, became the focus of vigorous research activity during the 1980s: Goodfellow et al. (1985) reviewed the Y chromsome, and Davies et al. (1987) presented the report of the Committee on the Genetic Constitution of the X and Y Chromosomes to the Ninth Human Gene Mapping Conference. Various forms of maps have been published (Vergnaud et al. 1986; Ferguson-Smith et al. 1987). The chromsome has merited a monograph (Sandberg 1985) and has become the subject of intense study (Goodfellow et al. 1987). However, few genes are known to be Y-linked in man, and there is little published information on the Y chromosome of other mammals, aside from the mouse. The existing data are presented below. An asterisk indicates that a probably homologous gene or region is known in man (Davies et al. 1987; McKusick 1986).

Molecular studies of the mouse Y chromosome

Eicher et al. (1983) and Eicher and Washburn (1986) described cytogenetic and molecular studies that resulted in partitioning of the normal Y chromosome into 6 functionally distinct regions containing the centromere; the *Tdy* and *Hya* loci, and *Bkm*-related sequences; sequences involved with sperm motility and with *Xmmv*; an X-pairing-and-recombinant segment; *Sts*; and the telomere. Bishop et al. (1985) used a Y-specific genomic DNA probe (pY353/B) to examine the RFLPs in 10 newly established mouse lines of the European semispecies, *Mus musculus domesticus* and *Mus musculus musculus*, and identified 2 variant forms of the Y chromosome, each characteristic of 1 of the semispecies. Avner et al. (1987) showed that 3 other probes – pY80/B, pY371/B, and pY302/B – isolated from a Y chromosome library, represent X–Y common sequences; using all 3 probes, they demonstrated exchange of genetic material between the X and Y chromosomes during meiosis.

AGGRESSIVE BEHAVIOR, INTERMALE

The Y-linked effect on intermale aggressive behavior between the DBA/1Bg and C57BL/10Bg strains and their hybrids described by Selmanoff et al.(1975, 1976) was confirmed by Maxson et al. (1979) in a more rigorously designed experiment. Selmanoff et al. (1977) suggested that this effect was correlated with a Y chromosome effect on developmental changes in serum testosterone. Stewart et al. (1980) found the same effect on aggression in crosses between 2 other strains, CBA/FaCam and C57BL/6Fa. The DBA/1, DBA/2, CBA, PHH, and Rb/1Bg Y chromosomes produce an increase in intermale aggression (Maxson et al. 1982; Shrenker and Maxson 1982).

AUTOIMMUNITY AND LYMPHOPROLIFERATION, ACCELERATED (*Yaa*)

Strain BXSB/Mp mice develop a spontaneous lupuslike syndrome which is strikingly accelerated in males (Andrews et al. 1978; Murphy and Roths 1978). When male BXSB mice are crossed with NZB/B1NJ, STL/J, or C57BL/6J females, their male offspring develop an accelerated autoimmune disease that does not occur in reciprocal hybrids or in females. The accelerated disease occurs only when the Y chromosome is derived from the recombinant inbred strain, BXSB, ultimately from strain SB/Le (Murphy and Roths 1978, 1979). The original investigations were extended by Theofilopoulos and Dixon (1981) and their colleagues and by Steinberg and his colleagues in a series of studies (see Steinberg et al. 1986 for summary). These studies indicated that there may be more than 1 factor on the SB/Le Y chromosome capable of accelerating immune disease, and that other non-Y genes are critical for the full expression of the disease. Izui et al. (1988) confirmed that *Yaa* by itself is not able to initiate autoimmune responses in mice that are not predisposed to autoimmune disease, and found that the gene seems able to initiate responses in mice that have the potential to develop autoimmune disease but whose autosomal abnormality is not sufficient to do so. Lahita et al. (1983) described several families in which systemic lupus erythematosus predominated in males. The disease appeared to be transmitted from affected males to their sons. The authors commented that the disease "had some similarity to the disease in BXSB mice."

BANDED KRAIT MINOR SATELLITE DNA (*Bkm*)

Certain nucleotide sequences from satellite DNA isolated from the W sex-determining chromosome of females of the elapid snake, *Bungarus fasciatus* (banded krait), are highly conserved in vertebrates, including mammals (see Jones 1983 for review). In mice, DNA related to these sequences is scattered throughout the male and female genomes, but is concentrated at the proximal tip of the Y chromosome (Jones and Singh 1981). This Y-linked DNA sequence consists of simple contiguous repeats of GATA tetranucleotides and appears to be transcribed in a sex-specific and developmentally regulated manner (Singh et al. 1984). The human Y chromosome does not contain clusters of these *Bkm*-related sequences (Kiel-Metzger et al. 1985).

CHEMOSENSORY IDENTITY

Yamazaki et al. (1986) demonstrated that genes located on the X and Y chromosomes determine chemosensory identity. See X-linked entry 30293 for details.

COMPETITIVE ABILITY OF SPERMATOZOA

Krzanowska (1986) provided evidence that Y-linked factors may influence the competitive ability of the spermatazoa.

DEFORMED SPERM HEADS

Males of the KE mouse strain have a high proportion (17%) of sperm with morphologically deformed heads. The trait is polygenically determined, and 1 of the responsible genes is on the Y chromosome (Krzanowska 1969, 1972).

GLUCOSE INTOLERANCE SYNDROME

When stressed by excessive handling and caging, nonobese males of the C3H.SW/SnJ mouse strain develop a syndrome characterized by intermittent mild hypoglycemia, extreme hyperinsulinemia, and extensive hyperplasia of the pancreatic beta cells (Leiter 1988). The impaired glucose tolerance can be circumvented by adrenolecto-mizing the males when they are weaned. When the Y chromosome from the C3HeB/FeChp background was transferred to the C3H.SW strain, the expression of the syndrome was reduced. Leiter suggests that environmental factors, mediated in part by the adrenal gland, interact with a genetic susceptibility determined by a gene on the Y chromosome to produce the impaired glucose tolerance.

*HISTOCOMPATIBILITY Y ANTIGEN (HY) (mouse, *Hya*)

The H–Y antigen was first described by Eichwald and Silmser (1955) as a male-specific minor transplantation antigen in the mouse. Billingham and Silvers (1959), Billingham et al. (1962), and subsequently others (Zeiss et al. 1962) described a male specific transplantation antigen in the rat. Welshons and Russell (1959) demonstrated that the Y chromosome of the mouse is male determining. The term "H–Y antigen" was first used by Billingham and Silvers (1960). Goldberg et al. (1971) developed a serological assay that was considered specific for the antigen, but this has never been conclusively proved (Silvers et al. 1982); this test has been used to demonstrate the extensive evolutionary conservation of the antigen. Silvers and Yang (1973) showed that the mouse and rat antigens are probably homologous, and Wachtel et al. (1974) found that the mouse H–Y antigen cross-reacts with those of the rabbit, guinea pig, rat, and man. The genetics and immunology of the antigen have been reviewed by Gasser and Silvers (1972), Wachtel (1977), and Simpson (1983). The finding that H–Y has been conserved over a long evolutionary period (Wachtel et al. 1975a) [Nakamura et al. (1987) provided additional evidence] led Wachtel et al. (1975b) to propose that the antigen might induce testicular development in mammals, and hence might be regarded as determining primary sex; Wachtel (1983) summarized the supporting evidence. However, the proposal was severely criticized by Fritz (1983) and Silvers et al. (1982). Silvers et al. (1982) suggested that H–Y be defined as a transplantation antigen, and that the serological entity be called the serologically detectable male (SDM) antigen. Nakamura et al. (1987) refer to the transplantation antigen as transplantation H–Y or H–Yt, and to the serological antigen as serological H–Y or H–Y. Brown and Migeon (1986) believe the methods of measuring the H–Y antigen and the SDM antigen define 2 different gene products. The discovery of a testis-determining locus in man and mouse (see entry in this Appendix) briefly complicated the picture further. However, it now seems clear that the locus determining the H–Y antigen and that responsible for the testis-determining factor are separate in man (Simpson et al. 1987a) and mouse (Simpson et al. 1987b). There is strong evidence suggesting that the H–Y antigen is the product of the gene for spermatogenesis (*Spy*) in the mouse (Burgoyne et al 1986; Burgoyne 1987). McLaren et al. (1988) and Roberts et al. (1988) independently demonstrated that *Hya* is on the minute short arm.

MOLONEY MURINE LEUKEMIA VIRUS INTEGRATION SITE-15 (M-MuLV INTEGRATION SITE-15) (*Mov-15*)

Münke et al. (1986: 148) noted in a table that this integration site is on both the Y and X chromosomes.

MOLONEY MURINE LEUKEMIA VIRUS INTEGRATION SITE-24 (M-MuLV INTEGRATION SITE-24) (*Mov-24*)

Münke et al. (1986: 148) noted in a table that this integration site is on the Y chromosome.

PSEUDOAUTOSOMAL SEQUENCES, HUMAN

Weber et al. (1987) demonstrated that a repeated element specific for the human pseudoautosomal region, *DXYZ2*, was present in the early replicating portion of the sex chromosomes of the chimpanzee, gorilla, and orangutan.

RESPONSE OF TARGET ORGANS TO TESTOSTERONE

Jutley and Stewart (1985) published evidence of 2 Y-linked loci influencing androgen metabolism. One affects the response of target organs, most significantly the seminal vescicles, to testosterone. At least 2 alleles exist, 1 in CBA/FaCamSt, the other in C57BL/FaSt.

SEROLOGICALLY DETECTABLE MALE (SDM) ANTIGEN

See MALE-SPECIFIC ANTIGEN.

SEX RATIO

Weir (1976) found that when male mice from the inbred strains PHH (sex ratio 0.535) and PHL (0.435) were mated to females of various inbred lines, the sex ratio followed the sire. This paternal effect, presumably due to the Y chromosome, persisted in progeny of F_2 males.

SEXUAL BEHAVIOR

Weir and Hogle (1973) and Weir (1976) found evidence of a Y-linked effect on differences in mating behavior between the PHH and PHL strains; Stewart et al. (1980) found no such effect, using C57BL/FaCam and C57BL/6Fa. Adams and Stewart (Stewart 1983) confirmed the original findings of Weir and Hogle, and Shrenker and Maxson (1984) found that the source of the Y chromosome affects the proportion of mice that mount in a test of male copulatory behavior.

*SPERMATOGENESIS (mouse, *Spy*)

Levy and Burgoyne (1986) found that XO germ cells are selectively eliminated from the spermatogenic epithelium of XO/XY and XO/XY/XYY mosaic mice, and concluded that the mouse Y chromosome carries a "spermatogenesis gene" (or genes) that acts autonomously in the germ cells. This gene (*Spy*), the gene for testes determination (*Tdy*), and the gene for the male-specific transplantation antigen (*Hya*), are contained in the *Sxr* fragment (Evans et al. 1982; Singh and Jones 1982). McLaren et al. (1984) found a variant *Sxr* (designated *Sxr'*) that carries the testis-determining information but does not confer H–Y antigenicity. Burgoyne et al. (1986) compared the testicular histology of XO *Sxr* and XO *Sxr'* males, and found that whereas there is active spermatogenesis, albeit abnormal, during the later stages in XO *Sxr* mice, it is virtually eliminated in XO *Sxr'* mice. The correlation of the loss of "spermatogenesis gene" function with the loss of H–Y antigenicity suggests that the transplantation antigen, H–Y, is the "spermatogenesis gene" product. Burgoyne (1987) summarized the role of the Y chromosome in spermatogenesis in the mouse, including the evidence that the H–Y antigen may be the molecule mediating *Spy* activity, and concluded that there may be another gene, separate from *Spy, Tdy,* and *Hya*, that is necessary for normal spermatogenesis.

Cryptorchism commonly occurs in certain sheep breeds (Dolling and Brooker 1964). The abdominal testes in the affected rams lack germ cells (Blackshaw and Samisoni 1967). Bruere et al. (1969) suggested that this deficiency is caused by a mutation at a Y-linked locus determining gonocyte formation.

STEROID SULFATASE

Keitges et al. (1985) found evidence for a functional Y-linked allele of the X-linked *Sts* locus in mice. See 31341 for details.

*TESTIS-DETERMINING FACTOR

Eicher et al. (1982) provided evidence for a testis-determining locus, *Tdy*, in the mouse. *Mus poschiavinus* carries a variant allele, *Tdy^{Pos}*, which, when transferred onto the C57BL/6J genome, causes XY individuals to develop as hermaphrodites or as females with 2 ovaries. The XY females are H–Y positive (Johnson et al. 1982; Simpson et al. 1983). The *M. domesticus* Y chromosome produces the same effect. McLaren et al. (1984) provided independent evidence for *Tdy*. McLaren et al. (1988) and Roberts et al. (1988) independently demonstrated that the locus is on the minute short arm. Burgoyne et al. (1988) found that Sertoli cells are exclusively XY in XX ↔ XY chimeric mouse testes, indicating that the differentiation of these cells is triggered by the cell-autonomous activity of *Tdy*. Page et al. (1987) cloned a 230-kb segment of the human Y chromosome that contains some or all of the testis-determining gene (*TDF*). Certain DNA sequences within this region have been highly conserved during evolution; apparently homologous sequences occur on the Y chromosome of several mammalian species examined. In particular, identical sequences are found within the *Tdy* region of the mouse. The nucleotide sequence of the conserved DNA in the human Y suggests that it encodes a protein with multiple "finger" domains. Very similar sequences also occur on the X chromosomes of humans and other mammals, and similar sequences occur in the chicken. German (1988) proposed that human gonadal dimorphism develops from the dosage effects of a locus (*GDL*) on the mammalian Y and X chromosomes.

TESTIS SIZE

Hayward and Shire (1974) found a Y-linked gene(s) that affects testis weight in adult C57BL/6 and CBA mice. The finding was confirmed by Herrick and Wolfe (1977) using several strains, and by Stewart et al. (1980) using strains CBA/FaCam and C57BL/6Fa. These studies indicate that the CBA and DBA/2 Y chromosomes produce a decrease in testis weight and an increase in intermale aggression, and that both traits may be influenced by the same Y chromosomal gene (Shrenker and Maxson 1986). Hunt and Mittwoch (1987) investigated testis size in 2 other inbred strains, BALB/c/Ola and CBA/Gr. The former has larger testes from day 14 of embryonic development. Testis size was affected by the origin of the Y chromosome, the X chromosome, the autosomes (or pseudoautosomal region, or both), and by maternal factors. McCoshen (1983) showed that the Y chromosome has a major effect on the rate of growth of gonadal somatic tissue in the mouse.

TESTOSTERONE LEVELS, SERUM

Selmanoff et al. (1977) suggested that the Y chromosome effect on intermale aggressive behavior in mice was correlated with a Y chromosome effect on developmental changes in serum testosterone. Jutley and Stewart (1985) described 2 Y-linked loci influencing androgen metabolism, one of which affects serum testosterone levels. At least 2 alleles exist: 1 in PHL/St, the other in PHL-YH/St.

VIRUS-RELATED SEQUENCES

Phillips et al. (1982) reported that the mouse Y chromosome contains sequences related to 2 different retroviruses. One of the probes used to identify the sequences also recognized a previously unidentified endogenous murine virus present in the Y chromosome; the virus is designated *MuRVY* (murine repeated virus on the Y chromosome) (unpublished data reported by Eicher and Washburn 1986: 345–346).

XENOTROPIC-MCF ENDOGENOUS VIRUS-Y (Xmmv-Y)

Blatt et al. (1983) showed that DNA sequences of the xenotropic-related *env* gene family, renamed *Xmmv* (Davisson 1986), in the mouse genome are on the Y chromosome.

X–Y UNIVALENCY, Y CHROMOSOME CONTROL OF

In the C57BL/6J-DBA/2J strain pair, X–Y univalency is controlled by 3 genetic systems, 1 of which appears to involve the Y chromosome (Biddle et al. 1985).

*Y CHROMOSOMAL SEQUENCES

Several groups (Nallaseth et al. 1983; Lamar and Palmer 1984; Bishop et al. 1985; Nallaseth and Dewey 1986; Nishioka and Lamothe 1986, 1987) have described Y chromosomal sequences isolated from mouse DNA. The male specificity is not absolute, and apparently homologous sequences have been found on the X chromosome (Nallaseth and Dewey 1986). Several studies (Koenig et al. 1984, 1985; Page et al. 1984; Erickson 1987) indicate that, although human Y chromosomal sequences are evolutionarily conserved among nonhuman primates, they are not located exclusively on the Y chromosome. Cotinot et al. (1987) cloned and characterized 2 bovine Y-specific DNA sequences, one of which appeared to be species specific.

References

Andrews BS, Eisenberg RE, Theofilopoulos AN, Izui S, Wilson CB, McConahey PJ, Murphy D, Roths JB, Dixon FJ: 1978. Spontaneous murine lupus-like syndromes: clinical and immunopathological manifestations in several strains. *J Exp Med* 148: 1198–1215.

Avner P, Bishop C, Amar L, Cambrou J, Hatat D, Arnaud D, Mattei M-G: 1987. Mapping the mouse X chromosome: possible symmetry in the location of a family of sequences on the mouse X and Y chromosomes. *Development* 101 (Suppl): 107–116.

Biddle FG, MacDonald BG, Eales BA: 1985. Genetic control of sex-chromosomal univalency in the spermatocytes of C57BL/6J and DBA/2J mice. *Can J Genet Cytol* 27: 741–750.

Billingham RE, Silvers WK: 1959. Inbred animals and tissue transplantation immunity. *Transplant Bull* 6: 399–403.

Billingham RE, Silvers WK: 1960. Studies on tolerance of the Y chromosome antigen in mice. *J Immunol* 85: 14–26.

Billingham RE, Hodge BA, Silvers WK: 1962. An estimate of the number of histocompatibility loci in the rat. *Proc Natl Acad Sci USA* 48: 138–147.

Bishop CE, Boursot P, Baron B, Bonhomme F, Hatat D: 1985. Most classical *Mus musculus domesticus* laboratory mouse strains carry a *Mus musculus musculus* Y chromosome. *Nature* 315: 70–72.

Blackshaw AW, Samisoni JI: 1967. The testes of the cryptorchid ram. *Res Vet Sci* 8: 187–194.

Blatt C, Mileham K, Hass M, Nesbitt MN, Harper ME, Simon MI: 1983. Chromosomal mapping of the mink cell focus-inducing and xenotropic *env* gene family in the mouse. *Proc Natl Acad Sci USA* 80: 6298–6302.

Brown TR, Migeon CJ: 1986. Androgen receptors in normal and abnormal male sexual differentiation. In: Chrousos GP, Loriaux DL, Lipsett MB (eds), *Steroid Hormone Resistance: Mechanisms and Clinical Aspects*. New York: Plenum Press, pp 227–255.

Bruere AN, McDonald, MF, Marshall RB: 1969. Cytogenetical analysis of an ovine male pseudohermaphrodite and the possible role of the Y chromosome in cryptorchidism of sheep. *Cytogenetics* 8: 148–157.

Burgoyne PS: 1987. The role of the mammalian Y chromsome in spermatogenesis. *Development* 101 (Suppl): 133–141.

Burgoyne PS, Levy ER, McLaren A: 1986. Spermatogenic failure in male mice lacking X–Y antigen. *Nature* 320: 170–172.

Burgoyne PS, Buehr M, Koopman P, Rossant J, McLaren A: 1988. Cell-autonomous action of the testis-determining gene: Sertoli cells are exclusively XY in XX ↔ XY chimaeric mouse testes. *Development* 102: 443–450.

Cotinot C, Kirszenbaum M, Bishop C, Leonard M, Vaiman M, Fellous M: 1987. Cloning and characterization of bovine Y derived sequences. (Abstr) *Cytogenet Cell Genet* 46: 598 only.

Davies KE, Mandel J-L, Weissenbach J, Fellous M: 1987. Report of the committee on the genetic constitution of the X and Y chromosomes. *Cytogenet Cell Genet* 46: 277–315.

Davisson M: 1986. Symbol changes for *Env* and *Xp* loci. *Mouse News Lett* 75: 5 only.

Dolling CHS, Brooker MG: 1964. Cryptorchidism in Australian merino sheep. *Nature* 203: 49–50.

Eicher EM, Washburn LL: 1986. Genetic control of primary sex determination in mice. *Annu Rev Genet* 20: 327–360.

Eicher EM, Washburn LL, Whitney JB III, Morrow KE: 1982. *Mus poschiavinus* Y chromosome in the C57BL/6J murine genome causes sex reversal. *Science* 217: 535–537.

Eicher EM, Phillips SJ, Washburn LL: 1983. The use of molecular probes and chromosomal rearrangements to partition the mouse Y chromosomes into functional regions. In: Messer A, Porter IH (eds), *Recombinant DNA and Medical Genetics.* New York: Academic Press, pp 57–71.

Eichwald EJ, Silmser CR: 1955. Untitled communication. *Transplant Bull* 2: 148–149.

Erickson RP: 1987. Evolution of four human Y chromosomal unique sequences. *J Mol Evol* 25: 300–307.

Evans EP, Burtenshaw MD, Cattanach BM: 1982. Meiotic crossing-over between the X and Y chromosomes of male mice carrying the sex-reversing (*Sxr*) factor. *Nature* 300: 443–445.

Ferguson-Smith MA, Affara NA, Magenis RE: 1987. Ordering of Y-specific sequences by deletion mapping and analysis of X–Y interchange males and females. *Development* 101 (Suppl): 41–50.

Fritz IB: 1983. The elusive H–Y antigen. (Book review) *Cell* 34: 1–2.

Gasser DL, Silvers WK: 1972. Genetics and immunology of sex-linked antigens. *Adv Immunol* 15:215–247.

German J: 1988. Gonadal dimorphism explained as a dosage effect of a locus on the sex chromosomes, the gonad-differentiation locus (*GDL*). *Am J Hum Genet* 42: 414–421.

Goldberg EH, Boyse EA, Bennett D, Scheid M, Carswell EA: 1971. Serological demonstration of H–Y (male) antigen on mouse sperm. *Nature* 232: 478–480.

Goodfellow P, Darling S, Wolfe J: 1985. The human Y chromosome. *J Med Genet* 22: 329–344.

Goodfellow PN, Craig IW, Smith JC, Wolfe J (eds): 1987. The mammalian Y chromosome: molecular search for the sex-determining factor. *Development* 101 (Suppl): 1–203.

Hayward P, Shire JGM: 1974. Y chromosome effect on adult testis size. *Nature* 250: 499–500.

Herrick CS, Wolfe HG: 1977. Effect of the Y-chromosome on testes size in the mouse (*Mus musculus*). (Abstr) *Genetics* 86: s27 only.

Hunt SE, Mittwoch U: 1987. Y-chromosomal and other factors in the development of testis size in mice. *Genet Res* 50: 205–211.

Izui S, Higaki M, Morrow D, Merino R: 1988. The Y chromosome from autoimmune BXSB/MpJ mice induces a lupus-like syndrome in the (NZW × C57BL/6) F_1 male mice, but not in C57BL/6 male mice. *Eur J Immunol* 18: 911–915.

Johnson LL, Sargent EL, Washburn LL, Eicher EM: 1982. XY female mice express H–Y antigen. *Dev Genet* 3: 247–253.

Jones KW: 1983. Evolutionary conservation of sex specific DNA sequences. *Differentiation* 23 (Suppl): S56–S59.

Jones KW, Singh L: 1981. Conserved repeated DNA sequences in vertebrate sex chromosomes. *Hum Genet* 58: 46–53.

Jutley JK, Stewart AD: 1985. Genetic analysis of the Y-chromosome of the mouse: evidence for two loci affecting androgen metabolism. *Genet Res* 47: 29–34.

Keitges E, Rivest M, Siniscalco M, Gartler SM: 1985. X-linkage of steroid sulphatase in the mouse is evidence for a functional Y-linked allele. *Nature* 315: 226–227.

Kiel-Metzger K, Warren G, Wilson GN, Erickson RP: 1985. Evidence that the human Y chromosome does not contain clustered DNA sequences (Bkm) associated with heterogametic sex determination in other vertebrates. *N Engl J Med* 313: 242–245.

Koenig M, Camerino G, Heilig R, Mandel J-L: 1984. A DNA fragment from the human X chromosome short arm which detects a partially homologous sequence on the Y chromosome's long arm. *Nucleic Acids Res* 12: 4097–4109.

Koenig M, Moison JP, Heilig R, Mandel J-L: 1985. Homologies between X and Y chromosomes detected by DNA probes: localisation and evolution. *Nucleic Acids Res* 13: 5485–5501.

Krzanowska H: 1969. Factor responsible for spermatozoan abnormality located in the Y chromosome in mice. *Genet Res* 13: 17–24.

Krzanowska H: 1972. Influence of Y chromosome on fertility in mice. In: Beatty RA, Gluecksohn-Waelsch S (eds), *The Genetics of the Spermatozoon.* Edinburgh: Department of Genetics of the University, pp 370–386.

Krzanowska H: 1986. Interstrain competition amongst mouse spermatozoa inseminated in various proportions, as affected by the genotype of the Y chromosome. *J Reprod Fertil* 77: 265–270.

Lahita RG, Chiorazzi N, Gibofsky A, Winchester RJ, Kunkel HG: 1983. Familial systemic lupus erythematosus in males. *Arthritis Rheum* 26: 39–44.

Lamar EE, Palmer E: 1984. Y-encoded, species-specific DNA in mice: evidence that the Y chromosome exists in two polymorphic forms in inbred strains. *Cell* 37: 171–177.

Leiter EH: 1988. Control of spontaneous glucose intolerance, hyperinsulinemia, and islet hyperplasia in nonobese C3H.SW male mice by Y-linked locus and adrenal gland. *Metabolism* 37: 689–696.

Levy ER, Burgoyne PS: 1986. The fate of XO germ cells in the testes of XO/XY and XO/XX/XYY mouse mosaics: evidence for a spermatogenesis gene on the mouse Y chromosome. *Cytogenet Cell Genet* 42: 208–213.

Maxson SC, Ginsburg BE, Trattner A: 1979. Interaction of Y-chromosomal and autosomal gene(s) in the development of intermale aggression in mice. *Behav Genet* 9: 219–226.

Maxson SC, Platt, T, Shrenker P, Trattner A: 1982. The influence of the Y-chromosome of Rb/1Bg mice on agonistic behaviors. *Agg Behav* 8: 285–291.

McCoshen JA: 1983. Quantitation of sex chromosomal influence(s) on the somatic growth of fetal gonads in vivo. *Am J Obstet Gynecol* 145: 469–473.

McKusick VA: 1986. *Mendelian Inheritance in Man: Catalogs of Autosomal Dominant, Autosomal Recessive, and X-Linked Phenotypes*, 7th ed. Baltimore: Johns Hopkins Univ Press, pp lxxv–lxxvi.

McLaren A, Simpson E, Tononari K, Chandler P, Hogg H: 1984. Male sexual differentiation in mice lacking H–Y antigen. *Nature* 312: 552–555.

McLaren A, Simpson E, Epplen JT, Studer R, Koopman P, Evans EP, Burgoyne PS: 1988. Location of the genes controlling H–Y antigen expression and testis determination on the mouse Y chromosome. *Proc Natl Acad Sci USA* 85: 6442–6445.

Münke M, Harbers K, Jaenisch R, Francke U: 1986. Chromosomal mapping of four different integration sites of Moloney murine leukemia virus including the locus for alpha 1 (I) collagen in mouse. *Cytogenet Cell Genet* 43: 140–149.

Murphy RD, Roths JB: 1978. Autoimmunity and lymphoproliferation: induction by mutant gene *lpr*, and acceleration by a male-associated factor in strain BXSB mice. In: Rose NR, Bigazzi PE, Warner NL (eds), *Genetic Control of Autoimmune Disease.* New York: Elsevier North Holland, pp 207–220.

Murphy ED, Roths JB: 1979. A Y chromosome associated factor in strain BXSB producing accelerated autoimmunity and lymphoproliferation. *Arthritis Rheum* 22: 1188–1194.

Nakamura D, Wachtel SS, Lance V, Becak W: 1987. On the evolution of sex determination. *Proc R Soc Lond [Biol]* 232: 159–180.

Nallaseth FS, Dewey MJ: 1986. Moderately repeated mouse Y chromosomal sequence families present distinct types of organization and evolutionary change. *Nucleic Acids Res* 14: 5295–5307.

Nallaseth FS, Lawther RP, Stallcup MR, Dewey MJ: 1983. Isolation of recombinant bacteriophage containing male-specific mouse DNA. *Mol Gen Genet* 190: 80–84.

Nishioka Y, Lamothe E: 1986. Isolation and characterization of a mouse Y chromosomal repetitive sequence. *Genetics* 113: 417–432.

Nishioka Y, Lamothe E: 1987. Evolution of a mouse Y chromosomal sequence flanked by highly repetitive elements. *Genome* 29: 380–383.

Page DC, Harper ME, Love J, Botstein D: 1984. Occurrence of a transposition from the X-chromosome long arm to the Y-chromosome short arm during human evolution. *Nature* 311: 119–123.

Page DC, Mosher R, Simpson EM, Fisher EMC, Mardon G, Pollack J, McGillivray B, de la Chapelle A, Brown LG: 1987. The sex-determining region of the human Y chromosome encodes a finger protein. *Cell* 51: 1091–1104.

Phillips SJ, Birkenmeier EH, Callahan R, Eicher EM: 1982. Male and female mouse DNAs can be discriminated using retroviral probes. *Nature* 297: 241–243.

Roberts C, Weith A, Passage E, Michot JL, Mattei MG, Bishop CE: 1988. Molecular and cytogenetic evidence for the location of *Tdy* and *Hya* on the mouse Y chromosome short arm. *Proc Natl Acad Sci USA* 85: 6446–6449.

Sandberg AA (ed): 1985. *The Y Chromosome*. New York: Alan R. Liss.

Selmanoff MK, Jumonville JE, Maxson SC, Ginsburg BE: 1975. Evidence for a Y chromosomal contribution to an aggressive phenotype in inbred mice. *Nature* 253: 529–530.

Selmanoff MK, Maxson SC, Ginsburg BE: 1976. Chromosomal determinants of intermale aggressive behavior in inbred mice. *Behav Genet* 6: 53–69.

Selmanoff MK, Goldman BD, Maxson SC, Ginsburg BE: 1977. Correlated effects of the Y-chromosome of mice on developmental changes in testosterone levels and intermale aggression. *Life Sci* 20: 359–365.

Shrenker P, Maxson SC: 1982. The Y chromosomes of DBA/1Bg and DBA/2Bg compared for effects on intermale aggression. *Behav Genet* 12: 429–434.

Shrenker P, Maxson SC: 1984. The DBA/1Bg and DBA/2Bg Y chromosomes compared for their effects on male sexual behavior. *Behav Neural Biol* 42: 33–37.

Shrenker P, Maxson SC: 1986. Effects of the DBA/1Bg Y chromosome on testis weight and aggression. *Behav Genet* 16: 263–270.

Silvers WK, Yang SL: 1973. Male specific antigen: homology in mice and rats. *Science* 181: 570–572.

Silvers WK, Gasser DL, Eicher EM: 1982. H–Y antigen, serologically detectable male antigen and sex determination. *Cell* 28: 439–440.

Simpson E: 1983. Immunology of H–Y antigen and its role in sex determination. *Proc R Soc Lond [Biol]* 220: 31–46.

Simpson E, Chandler P, Washburn LL, Bunker HP, Eicher EM: 1983. H–Y typing in karyotypically abnormal mice. *Differentiation* 23 (Suppl): S116–S120.

Simpson E, Chandler P, Goulmy E, Disteche CM, Ferguson-Smith MA, Page DC: 1987a.

Separation of the genetic loci for the H–Y antigen and for testis determination on human Y chromosome. *Nature* 326: 876–878.

Simpson E, Chandler P, McLaren A, Goulmy E, Disteche CM, Page DC, Ferguson-Smith MA: 1987b. Mapping the H–Y gene. *Development* 101 (Suppl): 157–161.

Singh L, Jones KW: 1982. Sex reversal in the mouse (*Mus musculus*) is caused by a recurrent nonreciprocal crossover involving the X and an aberrant Y chromosome. *Cell* 28: 205–216.

Singh L, Phillips C, Jones KW: 1984. The conserved nucleotide sequences of Bkm, which define *Sxr* in the mouse, are transcribed. *Cell* 36: 111–120.

Steinberg AD, Triem KH, Smith HR, Laskin CA, Rosenberg YJ, Klinman DM, Mushinski JF, Mountz JD: 1986. Studies of the effects of Y chromosome factors on the expression of autoimmune disease. *Ann NY Acad Sci* 475: 200–218.

Stewart AD: 1983. The role of the Y-chromosome in mammalian sexual differentiation. In: Johnson MH (ed), *Development in Mammals*. Amsterdam: Elsevier Science Publishers, p 340.

Stewart AD, Manning A, Batty J: 1980. Effects of Y-chromosome variants on the male behaviour of the mouse *Mus musculus*. *Genet Res* 35: 261–268.

Theofilopoulos AN, Dixon FJ: 1981. Etiopathogenesis of murine SLE. *Immunol Rev* 55: 179–216.

Vergnaud G, Page DC, Simmler M-C, Brown L, Rouyer F, Noel B, Botstein D, de la Chapelle A, Weissenbach J: 1986. A deletion map of the human Y chromosome based on DNA hybridization. *Am J Hum Genet* 38: 109–124.

Wachtel SS: 1977. H–Y antigen: genetics and serology. *Immunol Rev* 33: 33–58.

Wachtel SS: 1983. *H–Y Antigen and the Biology of Sex Determination*. New York: Grune & Stratton.

Wachtel SS, Koo GC, Zuckerman EE, Hammerling U, Scheid MP, Boyse EA: 1974. Serological crossreactivity between H–Y (male) antigens of mouse and man. *Proc Natl Acad Sci USA* 71: 1215–1218.

Wachtel SS, Koo GC, Boyse EA: 1975a. Evolutionary conservation of H–Y ('male') antigen. *Nature* 254: 270–272.

Wachtel SS, Ohno S, Koo GC, Boyse EA: 1975b. Possible role for X–Y antigen in the primary determination of sex. *Nature* 257: 235–236.

Weber B, Weissenbach J, Schempp W: 1987. Conservation of human-derived pseudoautosomal sequences on the sex chromosomes of the great apes. *Cytogenet Cell Genet* 45: 26–29.

Weir JH: 1976. Allosomal and autosomal control of sex ratio in PHH and PHL mice. *Genetics* 84: 755–764.

Weir JA, Hogle GA: 1973. Influence of the *Y* chromosome on sex ratio and mating behaviour in PHH and PHL mice. (Abstr) *Genetics* 74: s294 only.

Welshons WJ, Russell LB: 1959. The Y-chromosome as the bearer of male determining factors in the mouse. *Proc Natl Acad Sci USA* 45: 560–566.

Yamazaki K, Beauchamp GK, Matsuzaki O, Bard J, Thomas L, Boyse EA: 1986. Participation of the murine X and Y chromosomes in genetically determined chemosensory identity. *Proc Natl Acad Sci USA* 83: 4438–4440.

Zeiss IM, Nisbet NW, Heslop BF: 1962. A male antigen in rats. *Transplant Bull* 30: 161–163.

USING THE CATALOG

Because I am concerned with comparative aspects of the X chromosome, I have tried to create a companion volume to McKusick's catalog of X-linked phenotypes in man, *Mendelian Inheritance in Man* (MIM), and the format is patterned on it. I have used the same numbering system and have attempted to integrate the entries into McKusick's system: Apparently true homologies are given the number assigned in McKusick's catalog (with a few exceptions noted below); probable or possible homologies are given different numbers, but attention is drawn to the fact that a homologous relation with a human locus or phenotype might exist. To minimize any conflict or potential confusion, loci that are apparently unknown in man have been given numbers that are, as yet, unassigned in MIM. These general statements cover most entries; separate comment is required for a few others.

In attempting to comply with the recommendations of Lalley and McKusick (1985) noted in the Introduction, I have fitted homologous loci known in the mouse and other species into the human nomenclature system where appropriate; for example, the mottled series in the mouse and the mutation at the homologous locus in the Syrian hamster appear under MENKES SYNDROME. In one instance, although the same number (30150) is used, the entry is named after the normal product of the locus (Alpha-galactosidase A) not the mutant phenotype (Angiokeratoma, diffuse). Finally, in several instances, where I believe the MIM entry heading gives an inadequate notion of the nature of the locus, I have changed it. [This is a particularly important issue, because, as McKusick himself points out (MIM6: xvi), his catalogs "are in the last analysis listings of loci."] In these instances, most of which concern designations retained from earlier editions of MIM (see MIM6, p xii), the MIM number is given in a box immediately after the entry title.

Wherever possible, I have attempted to name a locus by its primary effect, function, or product, not by an allelic mutation, which, if known, is listed as a subentry. The most striking example of this is DIHYDROTESTOSTERONE RECEPTOR (30493), which appears in MIM under TESTICULAR FEMINIZATION SYNDROME (31370). A locus known only through the existence of an allelic mutation (i.e., the function of the "normal" allele is unknown) appears under the mutation. Because I have chosen, where possible, to follow the terminology and general format used in MIM, it has not always been possible to maintain this position. To make cross-referencing easier, allelic mutations that appear as subentries under their "normal" alleles, the original name of mutations that have been incorporated into the

Table 8. *Guide to an entry*

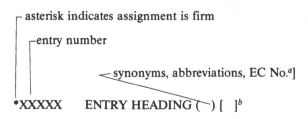

1. The entry may or may not have a general introduction containing information relevant to more than one species.
2. Information on each species is headed by the species' common name. The names are listed alphabetically. If the mutation has a name and abbreviation, they are given after the species name.
3. The references are given for the entire entry, not for individual species.

[a]Nomenclature Committee of the International Union of Biochemistry: 1984. *Recommendations of the Nomenclature Committee of the International Union of Biochemistry on the Nomenclature and Classification of Enzyme-Catalyzed Reactions*. Orlando FL: Academic Press.
[b]Square bracket, if present, contains:
1. MIM No., if the assigned entry number differs from number used in MIM for a homologous locus;
2. A "?" followed by MIM No(s). to indicate a locus that is probably or possibly homologous; or
3. "NK" indicating that the locus is unknown in man or, if known, is not yet an entry in MIM.

human nomenclature system, eponyms, and other alternative designations are entered in the subject index in such a way that they should be easily located in the catalog.

An asterisk placed before an entry number indicates that the assignment as an X-linked locus is secure. Absence of an asterisk indicates that the information available is not adequate to be sure that the locus is X-linked; there are 12 such entries (see Table 7).

The catalog is primarily a bibliographic guide; however, because it is impossible and unnecessary to provide a complete bibliography for all entries, I have attempted to select references that are historically relevant, are most current, or are particularly valuable because of the discussion included in them. I have tried to avoid citing veterinary case reports except in those instances where they make a significant contribution to an entry.

The manner in which each entry is set up is shown in Table 8.

Table 9. *Species commonly cited in catalog*

Common name	Scientific name
African green monkey	*Cercopithecus aethiops*
American mink	*Mustela vison*
Baboon	*Papio papio*
Capuchin monkey	*Cebus capucinus*
Cat	*Felus catus*
Chimpanzee	*Pan troglodytes*
Chinese hamster	*Cricetulus griseus*
Cattle	*Bos taurus*
Dasyrids	*Dasyuroidea*
Deer mouse	*Peromyscus maniculatus*
Dog	*Canis familiaris*
Gibbon	*Hylobates concolor*
Gorilla	*Gorilla gorilla*
Kangaroos and wallabies	*Macropodidae*
Horse	*Equus caballus*
Indian muntjac	*Muntiacus muntjac*
Mouse	*Mus musculus*
Mouse lemur	*Microcebus murinus*
Orangutan	*Pongo pygmaeus*
Owl monkey	*Aotus trivirgatus*
Pig	*Sus scrofa domestica*
Rabbit	*Oryctolagus cuniculus*
Rat[a]	*Rattus norvegicus*
Rhesus moneky	*Macaca mulatta*
Sheep	*Ovies ovies*
Syrian (golden) hamster	*Mesocricetus auratus*
Vole	*Microtus agrestis*
Wood lemming	*Myopus schisticolor*

[a]This is the brown (Norway) rat used as a laboratory animal. Occasionally the black rat (*Rattus rattus*) is the object of genetic study, and it will be distinguished by the adjective.

Species nomenclature

For convenience, I have used the common names for species mentioned in the catalog; the common and scientific names are given in Table 9. Occasionally, for information on loci or traits in rarer species, the scientific names are given in the relevant entry.

Conventions

The following abbreviations are used throughout: CNS, central nervous system; kb, kilobase; PNS, peripheral nervous system; RELP, restriction fragment length polymorphism.

No accents are used in headings. Greek letters are spelled out. Abbreviations used for genes are discussed in the Introduction.

CATALOG

30006 ABSENT PINNAE [NK]

Black rhinoceros

Absence of pinnae in the black rhinoceros has been recorded in at least 7 discrete populations in eastern and southern Africa (Goddard 1969). Of the 15 affected animals whose sex was known, 13 were males; 1 of the females may have been sired by an affected male, and the other was affected only unilaterally. Although relevant breeding data are virtually nonexistent, Goddard (1969) suggests that the trait may be X-linked. de Vos (1978) reported a unilaterally affected bull.

de Vos V: 1978. Congenital unilateral aotus in a black rhinoceros *Diceros bicornis bicornis* (Linn. 1758). *J S Afr Vet Assoc* 47: 71 only.

Goddard J: 1969. A note on the absence of pinnae in the black rhinoceros. *E Afr Wildl* 7: 178–181.

*30030 AGAMMAGLOBULINEMIA (BRUTON-TYPE AGAMMAGLOBULINEMIA; IMD1)

Cattle

Perk and Lobl (1962) studied a 3-month-old bull calf with agammaglobulinemia; the animal had repeated respiratory tract infections and several attacks of severe diarrhea, and died before additional immunologic studies could be carried out.

Horse

This disorder has been observed in Thoroughbred, Standardbred, and quarter horses. Four cases (all colts) were diagnosed among 2,516 horses evaluated for immunologic disorders (Perryman et al. 1983). Two of these have been described in detail (Banks et al. 1976; McGuire et al. 1976; Deem et al. 1979); both died at approximately $1^1/_2$ years. Pneumonia, enteritis, dermatitis, arthritis, and laminitis were noted at 2–6 months of age after the maternally derived immunoglobulins disappeared. B lymphocytes were absent, IgM and IgA could not be detected, and IgG and IgG(T) occurred in very low concentrations. The affected animals produced no detectable antibodies after immunization. T lymphocyte numbers and responses were normal. There were no lymphoid follicles or plasma cells in the lymph nodes and spleen, and the connective tissue stromal framework of the lymphoid follicles was absent. Perryman and McGuire (1980) briefly mentioned another probably affected animal that died at 2 months. Although the number of cases is small, the similarity of the disease to the X-linked form in man, and its occurrence in colts, makes the case for homology persuasive.

Mouse, X-linked immunodeficiency, *xid*

This mutation arose in a strain of mice at the National Institutes of Health, Bethesda (Scher et al. 1973). Hemizygotes and homozygotes have an immune deficiency characterized by a profound defect in B lymphocyte function; heterozygotes do not show any immunologic disorders. Affected animals have low serum IgM (Amsbaugh et al. 1972) and IgG3 (Perlmutter et al. 1979); do not respond to certain thymus-independent (TI) antigens (Amsbaugh et al. 1972; Scher et al. 1973), but do respond to others (Mond et al. 1978); and exhibit low responses to thymus-dependent (TD) antigens (Janeway and Barthold 1975; Scher et al. 1979). Early studies indicated that the primary error is in the number and function of B lymphocytes (Scher et al. 1975a,b), and subsequent studies indicated that a unique B cell population fails to develop (Scher et al. 1976). Subpopulations of mouse B cells express different amounts of 2 antigens, BLA-1 and BLA-2; Hardy et al. (1983, 1984) detected 2 B cell lineages, 1 of which (BLA-1$^+$, 2$^-$) is missing in CBA/N mice. Clayberger et al. (1985) presented evidence indicating that affected mice express defects in accessory and T cell subsets. The cellular basis of the defect is still unclear. The CBA/N strain, which "exists as a vigorous line that has no difficulty in dealing with the routine laboratory environment" (Scher 1981), has been used extensively in studies of the role of lymphocyte function in infections and autoimmune processes (see Scher 1982a for references). The locus is in the midregion of the chromosome (Berning et al. 1980), and occurs within or is very closely linked with an X-linked gene family, X-linked lymphocyte-regulated (XLR) (30906). XLR is present on late-stage B cell tumors and plasmacytomas (secretory B cell tumors), but cannot be detected in plasmacytomas carrying *xid* (Cohen et al. 1985). Dighiero et al. (1986) postulated that *xid* might be a regulatory gene because affected mice, unlike normal mice, are unable to secrete anti-DNA autoantibody on mitogenic stimulation, although they carry the genetic information to do so. Forrester et al. (1987) studied the effect of *xid* on subpopulations of B cells using *xid*/+ heterozygotes, and found that the mutation appeared to compromise the development of all B cells. Their data also suggested that, in competition with normal non-xid cells, newly formed xid B cells are less likely to be incorporated into the peripheral B cell pool. Miyake et al. (1987) described a defect in the migratory properties of B cells in CBA/N. Scher (1981, 1982a,b) and Wicker and Scher (1986) have published extensive reviews. The presence of several X-linked immunodeficiency diseases in man (MIM 30030, 30040, 30100, 30823, 30824) and of several loci involving immune responses in the mouse (30821, 30822, 30826, 30827, 30906) suggests that the mammalian X chromosome may contain an extended region controlling immune function; XLR (30906) may be the key to resolving the present uncertainties. Current evidence indicates that the *xid* mutation and that producing the Bruton-type agammaglobulinemia in mice are homologs.

Amsbaugh DF, Hansen CT, Prescott B, Stashak PW, Barthold DR, Baker PJ: 1972. Genetic control of the antibody response to type III pneumococcus polysaccharide in mice. I.

Evidence that an X-linked gene plays a decisive role in determining responsiveness. *J Exp Med* 136: 931–949.

Banks KL, McGuire TC, Jerrells TR: 1976. Absence of B lymphocytes in a horse with primary agammaglobulinemia. *Clin Immunol Immunopathol* 5: 282–290.

Berning AK, Eicher EM, Paul WE, Scher I: 1980. Mapping of the X-linked immune deficiency mutation (*xid*) of CBA/N mice. *J Immunol* 124: 1875–1877.

Clayberger C, DeKruyff RH, Fay R, Huber B, Cantor H: 1985. Evidence for defects in accessory and T cell subsets in mice expressing the *xid* defect. *J Mol Cell Immunol* 2: 61–68.

Cohen DI, Steinberg AD, Paul WE, Davis MM: 1985. Expression of an X-linked gene family (XLR) in late-stage B cells and its alteration by the *xid* mutation. *Nature* 314: 372–374.

Deem DA, Traver DS, Thacker HL, Perryman LE: 1979. Agammaglobulinemia in a horse. *J Am Vet Med Assoc* 175: 469–472.

Dighiero G, Poncet P, Rouyre S, Mazie J-C: 1986. Newborn xid mice carry the genetic information for the production of natural autoantibodies against DNA, cytoskeletal proteins, and TNP. *J Immunol* 136: 4000–4005.

Forrester LM, Ansell JD, Micklem HS: 1987. Development of B lymphocytes in mice heterozygous for the X-linked immunodeficiency (*xid*) mutation: *xid* inhibits development of all splenic and lymph node B cells at a stage subsequent to their initial formation in bone marrow. *J Exp Med* 165: 949–958.

Hardy RR, Hayakawa K, Parks DR, Herzenberg LA: 1983. Demonstration of B-cell maturation in X-linked immunodeficient mice by simultaneous three-color immunofluorescence. *Nature* 306: 270–272.

Hardy RR, Hayakawa K, Parks DR, Herzenberg LA, Herzenberg LA: 1984. Murine B cell differentiation lineages. *J Exp Med* 159: 1169–1188.

Janeway CA Jr, Barthold DR: 1975. An analysis of the defective response of CBA/N mice to T-dependent antigens. *J Immunol* 115: 898–900.

McGuire TC, Banks KL, Evans DR, Poppie MJ: 1976. Agammaglobulinemia in a horse with evidence of functional T lymphocytes. *Am J Vet Res* 37: 41–46.

Miyake K, Hayashi S-I, Ono S, Hamaoka T: 1987. The origin of antibody-forming cells detected in the bone marrow after thymus-independent antigen-stimulation and abnormality in migration of B cells of X-linked immunodeficient CBA/N mice. *Microbiol Immunol* 31: 821–829.

Mond JJ, Scher I, Mosier DE, Baese M, Paul WE: 1978. T-independent responses in B cell-defective CBA/N mice to *Brucella abortus* and to trinitrophenyl (TNP) conjugates of *Brucella abortus*. *Eur J Immunol* 8: 459–463.

Perk K, Lobl K: 1962. Agammaglobulinemia in a 3-month-old calf. *Am J Vet Res* 23: 171–173.

Perlmutter RM, Nahm M, Stein KE, Slack J, Zitron I, Paul WE, Davie JM: 1979. Immunoglobulin subclass-specific immunodeficiency in mice with an X-linked B-lymphocyte defect. *J Exp Med* 149: 993–998.

Perryman LE, McGuire TC: 1980. Evaluation for immune system failures in horses and ponies. *J Am Vet Med Assoc* 176: 1374–1377.

Perryman LE, McGuire TC, Banks KL: 1983. Infantile X-linked agammaglobulinemia: agammaglobulinemia in horses. *Am J Pathol* 111: 125–127.

Scher I: 1981. B-lymphocyte development and heterogeneity: analysis with the immune-defective CBA/N mouse strain. In: Gershwin ME, Merchant B (eds), *Immunologic Defects in Laboratory Animals*. New York: Plenum Press, pp 163–190.

Scher I: 1982a. CBA/N immune defective mice: evidence for the failure of a B cell subpopulation to be expressed. *Immunol Rev* 64: 117–136.

Scher I: 1982b. The CBA/N mouse strain: an experimental model illustrating the influence of the X-chromosome on immunity. *Adv Immunol* 33: 1–71.

Scher I, Frantz MM, Steinberg AD: 1973. The genetics of the immune response to a synthetic double-stranded RNA in a mutant CBA mouse strain. *J Immunol* 110: 1396–1401.

Scher I, Ahmed A, Strong DM, Steinberg AD, Paul WE: 1975a. X-linked B-lymphocyte immune defect in CBA/HN mice. I. Studies of the function and composition of spleen cells. *J Exp Med* 141: 788–803.

Scher I, Steinberg AD, Berning AK, Paul WE: 1975b. X-linked B-lymphocyte defect in CBA/N mice. II. Studies of the mechanisms underlying the immune defect. *J Exp Med* 142: 637–650.

Scher I, Sharrow SO, Paul WE: 1976. X-linked B-lymphocyte defect in CBA/N mice. III. Abnormal development of B-lymphocyte populations defined by their density of surface immunoglobulin. *J Exp Med* 144: 507–518.

Scher I, Berning AK, Asofsky R: 1979. X-linked B-lymphocyte defect in CBA/N mice. IV. Cellular and environmental influences on the thymus dependent IgG anti-sheep red blood cell response. *J Immunol* 123: 477–486.

Wicker LS, Scher I: 1986. X-linked immune deficiency (*xid*) of CBA/N mice. *Curr Top Microbiol Immunol* 124: 87–101.

30070 ALBINISM–DEAFNESS SYNDROME (ADFN)

Dog

Schaible (1982) suggested that Dalmatian dogs are a model of the human disorder; these dogs are frequently totally or partially deaf as early in life as their hearing can be tested, and most of them have completely unpigmented coats at birth. In both species, small pigmented spots develop all over the body postnatally. Complete or partial hypochromia also occurs frequently in Dalmatians. Schaible (1983) and Schaible et al. (1986) described additional details of the model; the sensorineural deafness appears to be a multifactorial trait correlated with elevated levels of uric acid in the urine. In none of these abstracts does Schaible comment directly on the critical point that the disorder is X-linked in man.

Schaible RH: 1982. Animal model of syndrome characterized by piebaldism and sensorineural deafness. (Abstr) *Am J Hum Genet* 34: 156A only.

Schaible RH: 1983. Sensorineural deafness is correlated with elevated level of uric acid in urine of animal model. (Abstr) *Am J Hum Genet* 35: 165A only.

Schaible RH, Cook JR Jr, Moorehead WR: 1986. Multifactorial inheritance with threshold effect in sensorineural deafness in animal model. (Abstr) *Am J Hum Genet* 39: A264 only.

*30150 ALPHA-GALACTOSIDASE (GLA; EC 3.2.1.22)

This is the structural locus for this enzyme.

American mink

Gradov et al. (1983) used somatic cell hybridization to demonstrate X-linkage.

Cat

O'Brien (1986) has shown that the locus is X-linked.

Cattle

Heuertz and Hors-Cayla (1978), Forster (1980), and Shimizu et al. (1981) demonstrated X-linkage by using cattle–mouse cell hybrids.

Dasyurids, *GLA*

Cooper et al. (1983) provided the first evidence for X-linkage in 3 species. Polymorphism exists and dosage compensation occurs through paternal X inactivation. In these marsupials the locus is syntenic with *G6PD, HPRT,* and *PGK* (Dobrovic and Graves 1986); in 2 species, *Planigale maculata* and *Sminthopsis crassicaudata,* the gene order appears to be *G6PD–HPRT–PGK–GLA,* with the first-named locus well separated from the others.

Dog

Meera Khan et al. (1984) used somatic cell hybridization to demonstrate X-linkage.

Macropodids (kangaroo and wallabies)

Cooper et al. (1983) demonstrated X-linkage in several species of macropodids. Polymorphism exists and dosage compensation occurs through paternal X-inactivation.

Mouse, *Ags*

The existence of an X-linked gene for this enzyme was first suggested by the results of cell hybridization experiments (Kozak et al., 1975). Lusis and West (1976) found a thermolabile variant in *M. musculus molossinus* and proved by genetic analysis that the structural gene was X-linked. They suggested *Ags^h* for the common allele present in most strains and *Ags^m* for the variant. Lusis and West (1978) provided evidence that the position of the *Ags* locus is distal to the *Mo* locus by about 9 centimorgans. The gene appears to be expressed in all adult tissues (Lusis and Paigen 1976), and is detectable in preimplantation embryos (Adler et al. 1977). Francke and Taggart (1980) used a mouse balanced reciprocal translocation in somatic cell hybridization experiments to show that the gene order in the mouse (centromere–*Hprt–Pgk–1–Ags*) is different from that in the human (centromere–*PGK–GAL–HPRT*). Francke et al. (1977) suggested that the locus should be assigned to the XE → XF region.

Pig

Forster (1980) and Leong et al. (1983) demonstrated X-linkage, using somatic cell hybrids.

Primates (nonhuman)

X-linkage has been demonstrated in the African green monkey, chimpanzee, gorilla, orangutan, and rhesus monkey (Garver et al. 1978), in the gibbon (Turleau et al. 1983), in the mouse lemur (Cochet et al. 1982), and in the owl monkey (Ma 1983).

Rabbit, *Gla*

Echard et al. (1981) showed that *Gla* is syntenic with *G6pd, Pgk,* and *Hprt.*

Rat

Levan et al. (1986) refer to unpublished work of Yoshida that supports X-linkage.

Red fox (*Vulpes vulpes*)

Rubtsov et al. (1987) demonstrated X-linkage.

Sheep

Saïdi-Mehtar (1981) demonstrated synteny with *G6pd* and *Pgk*, using somatic cell hybrids.

Virginia opossum (*Didelphis virginiana*)

The locus is probably X-linked in this species (Kaslow et al. 1987).

Adler DA, West JD, Chapman VM: 1977. Expression of alpha-galactosidase in preimplantation mouse embryos. *Nature* 267: 838–839.

Cochet C, Creau-Goldberg N, Turleau C, de Grouchy J: 1982. Gene mapping of *Microcebus murinus* (Lemuridae): a comparison with man and *Cebus capucinus* (Cebidae). *Cytogenet Cell Genet* 33: 213–221.

Cooper DW, Woolley PA, Maynes GM, Sherman FS, Poole WE: 1983. Studies on metatherian sex chromosomes. XII. Sex-linked inheritance and probable paternal *X*-inactivation of alpha-galactosidase A in Australian marsupials. *Aust J Biol Sci* 36: 511–517.

Dobrovic A, Graves JAM: 1986. Gene mapping in marsupials and monotremes. II. Assignments to the X chromosome of dasyurid marsupials. *Cytogenet Cell Genet* 41: 9–13.

Echard G, Gellin J, Benne F, Gillois M: 1981. The gene map of the rabbit (*Oryctolagus cuniculus* L.). I. Synteny between the rabbit gene loci coding for HPRT, PGK, G6PD, and GLA: their localization on the X chromosome. *Cytogenet Cell Genet* 29: 176–183.

Förster M: 1980. Localisation of X-linked genes in cattle and swine by somatic hybrid cells. In: *Fourth European Colloquium on Cytogenetics of Domestic Animals*. Uppsala: Swedish University of Agricultural Sciences, pp 322–331.

Francke U, Taggart RT: 1980. Comparative gene mapping: order of loci on the X-chromosome is different in mice and humans. *Proc Natl Acad Sci USA* 77: 3595–3599.

Francke U, Lalley PA, Moss W, Ivy J, Minna JD: 1977. Gene mapping in *Mus musculus* by interspecific cell hybridization: assignment of the genes for tripeptidase-1 to chromosome 10, dipeptidase-2 to chromosome 18, acid phosphatase-1 to chromosome 12, and adenylate kinase-1 to chromosome 2. *Cytogenet Cell Genet* 19: 57–84.

Garver JJ, Pearson PL, Estop A, Dijksman TM, Wijnen LMM, Westerveld A, Meera Khan P: 1978. Gene assignments to the presumptive homologs of human chromosomes 1, 6, 11, 12, and X in the Pongidae and Cercopithecoidea. *Cytogenet Cell Genet* 22: 564–569.

Gradov AA, Rubtsov NB, Shilov AG, Bochkarev MN, Serov OL: 1983. Chromosome localization of the genes for ENO1, HK1, ADK, ACP2, MPI, ITPA, ACON1 and alpha-GAL in the American mink (*Mustela vison*). *Theor Appl Genet* 67: 59–65.

Heuertz S, Hors-Cayla M-C: 1978. Carte génétique des bovins par la technique d'hybridation cellulaire. Localisation sur le chromosome X de la glucose-6-phosphate déshydrogénase, la phosphoglycérate kinase, l'alpha-galactosidase A et l'hypoxanthine–guanine phosphoribosyl transferase. *Ann Génét* 21: 197–202.

Kaslow DC, Migeon BR, Persico MG, Zollo M, VandeBerg JL, Samollow PB: 1987. Molecular studies of marsupial X chromosome reveal limited sequence homology of mammalian X-linked genes. *Genomics* 1: 19–28.

Kozak C, Nichols E, Ruddle RH: 1975. Gene linkage analysis in the mouse by somatic cell hybridization: assignment of adenine phosphoribosyltransferase to chromosome 8 and alpha-galactosidase to the X chromosome. *Somatic Cell Genet* 1: 371–382.

Leong MML, Lin CC, Ruth RF: 1983. The localization of genes for HPRT, G6PD, and alpha-GAL onto the X-chromosome of domestic pig (*Sus scrofa domesticus*). *Can J Genet Cytol* 25: 239–245.

Levan G, Szpirer J, Szpirer C, Yoshida MC: 1986. Present status of chromosome localization of rat genes. *Rat News Lett* 17: 3–8.

Lusis AJ, Paigen K: 1976. Properties of mouse alpha-galactosidase. *Biochim Biophys Acta* 437: 487–497.

Lusis AJ, West JD: 1976. X-linked inheritance of a structural gene for alpha-galactosidase in *Mus musculus*. *Biochem Genet* 14: 849–855.

Lusis J, West JD: 1978. X-linked and autosomal genes controlling mouse alpha-galactosidase expression. *Genetics* 88: 327–342.

Ma, NSF: 1983. Gene map of the new world Bolivian owl monkey, *Aotus*. *J Hered* 74: 27–33.

Meera Khan P, Brahe C, Wijnen LMM: 1984. Gene map of dog: six conserved and three disrupted syntenies. *Cytogenet Cell Genet* 37: 537–538.

O'Brien SJ (Frederick, Maryland): 1986. Personal communication, 86/12/23.

Rubtsov NB, Matveeva VG, Radjabli SI, Kulbakina NA, Nesterova TB, Zakian SM: 1987.

Construction of a clone panel of fox–Chinese hamster somatic cell hybrids and assignment of genes for LDHA, LDHB, GPI, ESD, G6PD, HPRT, alpha-GALA in the silver fox. (Russ, Eng summary) *Genetika* 23: 1088–1096.

Saïdi-Mehtar N, Hors-Cayla M-C, Van Cong N: 1981. Sheep gene mapping by somatic cell hybridization: four syntenic groups: ENO1–PGD, ME1–PGM3, LDHB–PEPB–TPI and G6PD–PGK–GALA. *Cytogenet Cell Genet* 30: 193–204.

Shimizu N, Shimizu Y, Kondo I, Woods C, Wenger T: 1981. The bovine genes for phosphoglycerate kinase, glucose-6-phosphate dehydrogenase, alpha-galactosidase, and hypoxanthine phosphoribosyltransferase are linked to the X chromosome in cattle–mouse hybrids. *Cytogenet Cell Genet* 29: 26–31.

Turleau C, Creau-Goldberg N, Cochet C, de Grouchy J: 1983. Gene mapping of the gibbon. Its position in primate evolution. *Hum Genet* 64: 65–72.

*30154 ANEMIA, X-LINKED [NK]

Mouse, sex-linked anemia, *sla*

This hypochromic, microcytic form of anemia, which regresses with age, was first observed among the male progeny of a female whose sire had been irradiated. X-linkage was demonstrated by Falconer and Isaacson (1962); the locus is in the midregion of the chromosome. The first general hematologic picture was presented by Grewal (1962), who stated that hemizygous males and homozygous females were easily recognized at birth by their pale color which persisted for only a few days. Affected animals are slightly smaller than their littermates at birth, but during the period of rapid postnatal growth they develop normally. However, after weaning they become anemic and their growth rate is reduced (Kingston and Bannerman 1974). Life expectancy appears to be normal. Aside from a few minor red cell abnormalities, heterozygous females are hematologically normal; however, their growth rate is also retarded, and, on an iron-deficient diet, they develop anemia much more rapidly than do normal female mice (Edwards et al. 1972). The studies of Bannerman and Pinkerton and their colleagues indicate that the anemia is caused by iron deficiency resulting from a defect in a genetically controlled step in the transfer of iron from the mucosal epithelial cells of the small intestine to the plasma (Bannerman and Cooper 1966; Pinkerton and Bannerman 1966, 1967; Bannerman and Pinkerton 1967; Pinkerton 1968; Pinkerton et al. 1970; Bedard et al. 1971). The intestinal defect cannot be demonstrated in newborns in which iron absorption is as effective as in normal littermates, but becomes apparent by the age of weaning (Kingston and Bannerman 1974). The effects of the defect can be modified by variations in diet (Kingston and Bannerman 1974; Sorbie et al. 1974). Studies in vitro using everted duodenal sacs (Edwards and Bannerman 1970; Manis 1970) and radioautographic studies with orally and intravenously administered iron (Bedard et al. 1973, 1976) demonstrated that the defect involves impaired transport of iron across the mucosa rather than impaired mucosal uptake. The defect, which is confined to the duodenal mucosa, is present to a diminished degree in heterozygous females (Manis 1971). The primary defect may be in the regulation of the synthesis of mucosal transferrin or the binding of iron to this protein (Heubers et al. 1973; Edwards and Hoke 1978). Studies of the kinetics of iron uptake by isolated intestinal cells suggest that the site of the error may be between the mucosal cell and the vascular compartment rather than in the cell (Peppriell et al. 1982). Grewal's original observation that affected animals were anemic at birth suggested that placental iron transfer is also affected by *sla*. Dancis and Jansen (1970) found that affected males are anemic in utero just before birth but could not demonstrate a defect in the delivery of iron to these fetuses in acute experiments. However, Kingston and Bannerman (1974) and Kingston et al. (1978) found that the placental transfer of radioiron is lower in affected mice than in their littermate controls when the radioiron is administered continuously to the dams. These findings indicate that the *sla* gene also produces a defect in placental iron transport. An X-linked hypochromic anemia, sideroblastic anemia (ASB) (MIM 30130), is known in man. Although the phenotypic features produced by *ASB* in man and *sla* in the mouse differ significantly, the locus in each species is located close to the respective locus determining G6PD.

Bannerman RM, Cooper RG: 1966. Sex-linked anemia; a hypochromic anemia of mice. *Science* 151: 581–582.

Bannerman RM, Pinkerton PH: 1967. X-linked hypochromic anaemia of mice. *Br J Haematol* 13: 1000–1013.

Bedard YC, Pinkerton PH, Simon GT: 1971. Ultrastructure of duodenal mucosa of mice with a hereditary defect of iron absorption. *J Pathol* 104: 45–51.

Bedard YC, Pinkerton PH, Simon GT: 1973.Radioautographic observations on iron absorption by the duodenum of mice with iron overload, iron deficiency, and X-linked anemia. *Blood* 42: 131–140.

Bedard YC, Pinkerton PH, Simon GT: 1976. Uptake of circulating iron by the duodenum of normal mice and mice with altered iron stores, including sex-linked anemia: high resolution radioautographic study. *Lab Invest* 34: 611–615.

Dancis J, Jansen V: 1970. Placental transport of iron in X-linked anaemia of mice. *Br J Haematol* 19: 573–578.

Edwards JA, Bannerman RM: 1970. Hereditary defect of intestinal iron transport in mice with sex-linked anemia. *J Clin Invest* 49: 1869–1871.

Edwards JA, Hoke JE: 1978. Mucosal iron binding proteins in sex-linked anemia and microcytic anemia of the mouse. *J Med* 9: 353–364.

Edwards JA, Bannerman RM, Kreimer-Birnbaum M: 1972. Sex-linked anemia of the mouse: an effect on the gene in the heterozygous state. (Abstr) *XV Int Congr Hematol Brazil.* Abstr 12.

Falconer DS, Isaacson JH: 1962. The genetics of sex-linked anaemia in the mouse. *Genet Res* 3: 248–250.

Grewal MS: 1962. A sex-linked anaemia in the mouse. *Genet Res* 3: 238–247.

Huebers H, Huebers E, Forth W, Rummel W: 1973. Iron absorption and iron-binding proteins in intestinal mucosa of mice with sex linked anaemia. *Hoppe Seylers Z Physiol Chem* 354: 1156–1158.

Kingston PJ, Bannerman RM: 1974. Iron metabolism in X-linked anaemia (*sla*): effects of the gene during development. (Abstr) *Br J Haematol* 27: 360–361.

Kingston PJ, Bannerman CEM, Bannerman RM: 1978. Iron deficiency anaemia in newborn *sla* mice: a genetic defect of placental iron transport. *Br J Haematol* 40: 265–276.

Manis J: 1970. Active transport of iron by intestine: selective genetic defect in the mouse. *Nature* 227: 385–386.

Manis J: 1971. Intestinal iron-transport defect in the mouse with sex-linked anemia. *Am J Physiol* 220: 135–139.

Peppriell JE, Edwards JA, Bannerman RM: 1982. The kinetics of iron uptake by isolated intestinal cells from normal mice and mice with sex-linked anemia. *Blood* 60: 635–641.

Pinkerton PH: 1968. Histological evidence of disordered iron transport in the X-linked hypochromic anaemia of mice. *J Pathol Bacteriol* 95: 155–165.

Pinkerton PH, Bannerman RM: 1966. X-linked hypochromic anemia of mice – a disorder of iron metabolism. (Abstr) *Blood* 28: 987 only.

Pinkerton PH, Bannerman RM: 1967. Hereditary defect of iron absorption in mice. *Nature* 216: 482–483.

Pinkerton PH, Bannerman RM, Doeblin TD, Benisch BM, Edwards JA: 1970. Iron metabolism and absorption studies in the X-linked anaemia of mice. *Br J Haematol* 18: 211–228.

Sorbie J, Hamilton DL, Valberg LS: 1974. Effect of various factors on iron absorption in mice with X-linked anaemia. *Br J Haematol* 25: 559–569.

*30179 ARAF PROTOONCOGENE [NK]

The cells of various vertebrates contain DNA sequences that are apparently homologous to viral oncogenes (v-oncs) carried by transforming retroviruses. (DNA sequences that are apparently homologous to vertebrate oncogenes occur in *Drosophila melanogaster*; see Shilo and Weinberg 1981.) These protooncogenes or cellular oncogenes (c-oncs) can be activated by retroviral transduction, mutation, or rearrangement to become retroviral oncogenes or cellular transforming genes (Hunter 1984). Evidence indicates that the v-oncs arose from the c-oncs by recombinational events (Bishop 1983; Duesborg 1983). The fact that the c-oncs have been conserved over a long evolutionary time period suggests that they play a vital role in normal development; the nature of this role has not been determined.

Mouse, Raf-related oncogene (*Araf*)

Huebner et al. (1986) isolated mouse and human cDNAs, designated mA-*raf* and hA-*raf*, respectively, because they are related to but distinct from c-*raf-1*, and used them to show that mA-*raf* and hA-*raf* are located on the respective X chromosomes. The hA-*raf-1* gene is located in the Xp21–Xq11 interval. Because the mA-*raf* is located 10–17 centimorgans proximal to *Hprt*, Avner et al. (1987) suggest that hA-*raf* is located on the short arm.

Avner P, Bucan M, Arnaud D, Lehrach H, Rapp U: 1987. A-*raf* oncogene localizes on mouse X chromosome to region some 10–17 centimorgans proximal to hypoxanthine phosphoribosyltransferase gene. *Somatic Cell Mol Genet* 13: 267–272.

Bishop JM: 1983. Cellular oncogenes and retroviruses. *Annu Rev Biochem* 52: 301–354.

Duesborg PH: 1983. Retroviral transforming genes in normal cells? *Nature* 304: 219–226.

Huebner K, Ar-Rushdi A, Griffen CA, Isobe M, Kozak C, Emanuel BS, Nagarajan L, Cleveland JL, Bonner TI, Goldsborough MD, Croce CM, Rapp U: 1986. Actively transcribed genes in the *raf* oncogene group, located on the X chromosome in mouse and man. *Proc Natl Acad Sci USA* 83: 3934–3938.

Hunter T: 1984. Oncogenes and proto-oncogenes: how do they differ? *J Natl Cancer Inst* 73: 773–786.

Shilo B-Z, Weinberg RA: 1981. DNA sequences homologous to vertebrate oncogenes are conserved in *Drosophila melanogaster. Proc Natl Acad Sci USA* 78: 6789–6792.

*30184 BALB RETROVIRUS RESTRICTION I [NK]

Cat, *BVR1*

O'Brien (1976) described the detection of this feline restriction gene that acts dominantly to block an endogenous B-tropic murine leukemia virus in cat–mouse hybrid cells. The gene is linked to the structural genes for HPRT and G6PD.

O'Brien SJ: 1976. *Bvr-1*, a restriction locus of a type C RNA virus in the feline cellular genome: identification, location, and phenotypic characterization in cat × mouse somatic cell hybrids. *Proc Natl Acad Sci USA* 73: 4618–4622.

*30188 BENT TAIL [NK]

Mouse, *Bn*

Garber (1952a,b) described this mutation that affects the structure and length of the tail. Hemizygous males and homozygous females have tails about $1/2$ normal length with several definite kinks or bends. Heterozygous females have normal-length tails but exhibit variable expressivity in the structural defect that may range from a single almost imperceptible bend to a tail curled and pressed against the body. Grüneberg (1955) presented more details on the nature of the skeletal defects. The anomalies arise during the formation of the tail on the 11th and 12th days of embryogenesis. Penetrance is 100% in hemizygous males and homozygous females. In Garber's original stock, penetrance in heterozygous females was at least 95%; however, Falconer (1954) and Grüneberg (1955) found about 15% normal overlaps. Half the affected males and all the homozygous females are smaller at birth and mature more slowly than normal mice. The original breeding data suggest that heterozygous females are less fertile than normal females, and homozygous females are less fertile than heterozygous females. In addition, there is a suggestion of a lethal effect of the gene in males and homozygous females. These observations indicate that the gene may have a more generalized systemic effect. Garber (1952b) described 4 affected males that "died within a few months" of birth and in which there was a flexion deformity of the forefeet. Hemizygotes and homozygotes often have an open neural tube in the sacral region and cranioschisis in utero (Butler and Lyon, quoted as a personal communication by Johnson 1976); adult animals often have an interfrontal bone and a slightly wider than normal skull.

Falconer DS: 1954. Linkage in the mouse: the sex-linked genes and "rough." *Z Indukt Abstammungs-Verebungsl* 86: 263–268.
Garber ED: 1952b. "Bent-tail," a dominant, sex-linked mutation in the mouse. *Proc Natl Acad Sci USA* 38: 876–879.
Grüneberg H: 1955. Genetical studies on the skeleton of the mouse. XVII. Bent-tail. *J Genet* 53: 551–562.
Johnson DR: 1976. The interfrontal bone and mutant genes in the mouse. *J Anat* 121: 507–513.

*30196 BROAD HEADED [NK]

Mouse, *Bhd*

The first mutant with this disorder occurred among the offspring of an irradiated female. The original male had a broad head and nose; he produced 10 female offspring with the same phenotype and 13 normal males. Aside from the first male, no affected males have lived to weaning; the trait may be partially lethal in females as well. The locus is close to *Ta* (Phillips and Fisher 1978).

Phillips RJS, Fisher G: 1978. Private communication. *Mouse News Lett* 58: 43–44.

*30214 CARPAL SUBLUXATION [NK]

Dog

Pick et al. (1967) described this condition, the locus of which is closely linked to that for hemophilia A. The defect is limited to the carporadial joints and always occurs bilaterally. No other anomaly is present. McKusick (1978) points out that, although there is no human homology, the occurrence of this defect is of great interest in view of the extensive homology in the mammalian X chromosome.

McKusick VA: 1978. *Mendelian Inheritance in Man: Catalogs of Autosomal Dominant, Autosomal Recessive, and X-Linked Phenotypes*, 5th ed. Baltimore: Johns Hopkins Univ Press, p lxxxviii.
Pick JR, Goyer RA, Graham JB, Renwick JH: 1967. Subluxation of the carpus in dogs: an X chromosomal defect closely linked with the locus for hemophilia A. *Lab Invest* 17: 243–248.

*30221 CATARACT, X-LINKED [?30220]

Mouse, *Xcat*

Favor and Pretsch (1987) briefly described the effects of this mutation, which was produced by paternal radiation. Hemizygous males and homozygous females have total lens opacity. Heterozygotes have variable expressivity ranging from clear lenses to total opacity. There are no associated eye, ear, or coat color effects. The locus is situated midway between *Plp* (31208) and *Hyp* (30780). The description given does not permit any definite decision regarding homology with the 1 form of human X-linked cataract that could be considered a candidate.

Favor J, Pretsch W: 1987. Position of *Xcat*, a new X-linked cataract mutation. *Mouse News Lett* 77: 139 only.

*30234 cDNA pMIF3/10 INDUCED WITH INTERFERON [NK]

Mouse

Skup et al. (1982) coinduced 2 incomplete murine cDNAs with interferon (IFN) in mouse cells infected with Newcastle disease virus and identified them as IFN genes; however, neither is very similar to the mouse IFN-alpha and -beta genes. Kelly et al. (1986) showed that one of these, pMIF3/10, is X-linked.

Kelley KA, Pitha PM, De Maeyer-Guignard J, De Maeyer E, Kozak C: 1986. Assignment of two mouse genes coinduced with interferon to chromosomes 12 and X. *J Interferon Res* 6: 51–57.
Skup D, Windass JD, Sor F, George H, Williams BRG, Fukuhara H, De Maeyer-Guignard J, De Maeyer E: 1982. Molecular cloning of partial cDNA copies of two distinct IFN-beta mRNAs. *Nucleic Acids Res* 10: 3069–3084.

*30236 CELL SURFACE ANTIGEN VP382 [NK]

Cat

VedBrat et al. (1980) described how a panel of monoclonal antibodies prepared against antigens expressed in feline lymphoma cells detect antigens (VP) present in normal mature lymphocytes and fibroblasts. The expression of one of these, VP382, is correlated with that of HPRT and G6PD (VedBrat et al. 1983). None of the known human X-linked antigens correlates with the expression of VP382.

VedBrat SS, Hammerling U, Hardy WD Jr, Borenfreund E, Prensky W: 1980. Monoclonal antibodies: detection of transformation-related antigens. *Cold Spring Harbor Symp Quant Biol* 44: 715–720.
VedBrat SS, Yu LC, Hammerling U, Prensky W: 1983. Detection of a feline X-linked antigen in somatic cell hybrids: single-cell analysis using monoclonal antibodies. *J Hered* 74: 75–80.

*30237 CELL SURFACE ANTIGEN, X-LINKED (SA-X) [?31345, 31346, 31347]

Red kangaroo (*Macropus rufus*)

Sykes and Hope (1978) used marsupial–eutherian somatic cell hybrids to detect an X-linked cell surface antigen in *M. rufus*. The relation of this antigen to X-linked surface antigens in the human is unknown.

Sykes PJ, Hope RM: 1978. The use of marsupial × eutherian somatic cell hybrids to study marsupial cell surface antigens. *Aust J Exp Biol Med Sci* 56: 703–711.

*30293 CHEMOSENSORY IDENTITY [NK]

Mouse

Each mouse possesses an odor reflecting its genetic constitution at the major histocompatibility complex on chromosome 17 (Yamazaki et al. 1976). Yamazaki et al. (1986) tested mice differing only in their X or Y chromosomes or in both for individuality of scent in a Y-maze system previously used to investigate the scent distinctions related to the major histocompatibility complex. They demonstrated that genes located on the X and Y chromosomes determine chemosensory identity.

Yamazaki K, Boyse EA, Mike V, Thaler HT, Mathieson BJ, Abbott J, Boyse J, Zayas ZA, Thomas L: 1976. Control of mating preferences in mice by genes in the major histocompatibility complex. *J Exp Med* 144: 1324–1335.
Yamazaki K, Beauchamp GK, Matsuzaki O, Bard J, Thomas L, Boyse EA: 1986. Participation of the murine X and Y chromosomes in genetically determined chemosensory identity. *Proc Natl Acad Sci USA* 83: 4438–4440.

*30295 CHONDRODYSPLASIA PUNCTATA, X-LINKED (CDPX)

Mouse, bare patches, *Bpa*

This mutation results in a semidominant male-lethal disorder (Phillips et al. 1973). At about 5 days of age, heterozygous females exhibit patches of bare skin among the first emerging hairs. Several days later, scurf flakes and occasionally scabs form around the patches. The scabs may last several weeks, but by 12 days, most of the patches are overlapped by normal growing hair, and the animals appear similar to striated (31351) and tabby (30510) heterozygotes. Many heterozygotes are smaller and weigh less than their normal sibs. Many also have hind limb abnormalities in which the toes are bent and sometimes shortened. In a few cases, the forefeet are affected and other skeletal anomalies may occur. The formation of scabs may produce abnormal growth of the ears and tail. The more severe the coat defect, the more likely the other defects are to occur. Hemizygous males probably die at the small mole stage of embryogenesis. About half the expected number of heterozygous females is observed.

The remainder appear to lose their mutant chromosome and become XO females. The high production of XO offspring is not due to *Bpa* itself but to a closely linked factor (Phillips and Kaufman 1974) that proved to be a long inversion (Evans and Phillips 1975). Happle (1979) suggested homology with the X-linked dominant form of chondrodysplasia punctata (MIM 30295), and detailed comparative studies support this view (Happle et al. 1983). Both traits behave as X-linked semidominant lethals, and heterozygous females exhibit strikingly similar types of skeletal dysplasia, including punctate enchondral calcifications, cataracts, ichthyosis, patchy and streaked hairlessness, and linearly patterned hyperpigmentation. However, Yang-Feng et al. (1986) point out that the relative map positions of the 2 loci differ (p 8683), and Davisson (1987) suggests that the mouse locus may be homologous with dyskeratosis congenita (MIM 30500).

Davisson MT: 1987. X-linked genetic homologies between mouse and man. *Genomics* 1: 213–227.

Evans EP, Phillips RJS: 1975. Inversion heterozygosity and the origin of XO daughters of *Bpa/+* female mice. *Nature* 256: 40–41.

Happle R: 1979. X-linked dominant chrondrodysplasia punctata: review of literature and report of a case. *Hum Genet* 53: 65–73.

Happle R, Phillips RJS, Roessner A, Junemann G: 1983. Homologous genes for X-linked chondrodysplasia punctata in man and mouse. *Hum Genet* 63: 24–27.

Phillips RJS, Kaufman MN: 1974. Bare patches, a new sex-linked gene in the mouse, associated with a high production of XO females. II. Investigations into the nature and mechanisms of the XO production. *Genet Res* 24: 27–41.

Phillips RJS, Hawker SG, Moseley HJ: 1973. Bare patches, a new sex-linked gene in the mouse, associated with a high production of XO females. I. A preliminary report of breeding experiments. *Genet Res* 22: 91–99.

Yang-Feng TL, De Gennaro LJ, Francke U: 1986. Genes for synapsis I, a neuronal phosphoprotein, map to conserved regions of human and murine X chromosomes. *Proc Natl Acad Sci USA* 83: 8679–8683.

*30336 CLEFT PALATE, CORTISONE-INDUCED, LIABILITY TO [NK]

Mouse

Strains differ in their sensitivity to cleft palate induction by cortisone and other corticoids (review, Juriloff 1980). In studies using the A/J, C57BL/6J, and C3H/HeJ strains, Francis (1973) demonstrated that X-linked loci influence the liability to cortisone-induced cleft palate.

Francis BM: 1973. Influence of sex-linked genes on embryonic sensitivity to cortisone in three strains of mice. *Teratology* 7: 119–226.

Juriloff DM: 1980. The genetics of clefting in the mouse. In: Melnick M, Bixler D, Shields ED (eds), *Etiology of Cleft Lip and Cleft Palate*. New York: Alan R. Liss, pp 39–71.

*30337 CLEFT PALATE, POLYDACTYLY, SYNDACTYLY, TIBIOFIBULAR SHORTENING, BRACHYGNATHISM, AND SCOLIOSIS [NK]

Dog

Sponenberg and Bowling (1985) described this syndrome of skeletal defects in a family of Australian shepherd dogs. The complete syndrome, found only in males, comprises cleft palate, syndactyly, polydactyly, tibiofibular shortening, brachygnathism, and often scoliosis; all affected males die shortly after birth. Affected females lack the cleft palate, brachygnathism, and scoliosis; in general, the expressed defects are so mild that an affected female could be overlooked in the absence of progeny data. The breeding data suggest the trait is an X-linked male lethal, but the possibility that it is a sex-influenced autosomal trait cannot be ruled out. Several X-linked syndromes of which cleft palate is a part are listed in MIM; however, polydactyly is described in none of them, and syndactyly is part of the oral–facial–digital syndrome (MIM 31120) only.

McKusick VA: 1986. *Mendelian Inheritance in Man: Catalogs of Autosomal Dominant, Autosomal Recessive, and X-linked Phenotypes*, 7th ed. Baltimore: Johns Hopkins Univ Press.

Sponenberg DP, Bowling AT: 1985. Heritable syndrome of skeletal defects in a family of Australian shepherd dogs. *J Hered* 76: 393–394.

*30356 COAGULATION FACTOR VIII (HEMOPHILIA A; CLASSICAL HEMOPHILIA; F8C) [30670]

Cat

Cotter et al. (1978) described 3 unrelated male cats with factor VIII deficiency. The mutation has been recognized 9 times in the United States. It is maintained in the domestic shorthair and crossbreeds in Albany, New York (Dodds 1981).

Cattle

Healy et al. (1984) investigated the cause of mortality, within 18 hours of castration, among Hereford calves on 4 farms in southern New South Wales. Three calves studied before castration all had extended whole blood clotting times, and one had an extended partial thromboplastin time; all 3 died as a result of hemorrhage after castration. Another calf with extended whole blood clotting and partial thromboplastin times had severely reduced levels (8%) of factor VIII. Unfortunately, no detailed breeding records were maintained for any of the herds, and the mode of transmission remains to be proved.

Dog

Early reports on "hemophilia" in the dog predate an adequate understanding of coagulation mechanisms and lack detailed pedigree data. The first extensive detailed studies were those of Field et al. (1946), Hutt et al. (1948), and Graham et al. (1949) on a colony of Irish setters. The defect occurs in most breeds (Dodds 1974; Graham et al. 1975). Clinical expression varies from mild to severe with the degree of factor VIII deficiency. Hemorrhages occur at numerous sites and seem to be more frequent in larger breeds. Affected animals have very low factor VIII coagulant activity but normal or elevated levels of factor VIII antigen; in carrier females, factor VIII levels are about 50% of normal (Parks et al. 1964). The disease has been described in female homozygotes created by crossing hemizygous males with heterozygous females (Brinkhous and Graham 1950). Only 1 case of spontaneous hemophilia in a bitch has been reported (Murtaugh and Dodds 1988); unfortunately, the animal was a crossbreed whose parents were not available for study. Brinkhous et al. (1973) bred 2 strains, each segregating for a different type of hemophilia, and found that the genes recombined freely. Double heterozygotes in repulsion for both A and B could be identified readily. When both hemophilia genes were in the coupling phase, there was evidence of increased intrauterine or neonatal lethality in males. Human and canine factor VIII are structurally similar (Bouma et al. 1976), and Benson and Dodds (1976) provided evidence for separate antigenic and precoagulant sites in canine factor VIII, and for heterogeneity of the antigenic determinants. There is considerable evidence that the human and canine diseases are true homologs (Brinkhouse and Griggs 1979); the advent of modern therapy in the human disorder developed directly from studies on hemophilic dogs. The characteristics and availability of colonies of hemophilic dogs in the United States, England, and the Netherlands are listed by Dodds (1981). Spurling (1980) reviewed hemophilia A and other hereditary canine disorders of hemostasis, and Dodds (1982) described a mass-screening program to detect inherited bleeding disorders.

Horse

The first case of hemophilia in the horse was reported by Archer (1961) and Nossel et al. (1962); unfortunately, no pedigree data were presented. Sanger et al. (1964) reported the first case in which there was a familial history: The dam of the affected colt had foaled 3 other colts who had died from hemorrhage; her fillies were normal. Hutchins et al. (1967) reported a case from Australia. Archer and Allen (1972) described 2 additional cases related to the one described by Archer in 1961; they were half brothers with a common dam, the half sister of the original case.

Mouse, *Cf-8*

Mullins et al. (1987) state that the gene in the mouse is located in the midpart of the chromosome near *G6pd* and *Rsvp*. No mutations at the locus are known.

Archer RK: 1961. True haemophilia (haemophilia A) in a Thoroughbred foal. *Vet Rec* 73: 338–340.

Archer RK, Allen BS: 1972. True haemophilia in horses. *Vet Rec* 91: 655–656.

Benson RE, Dodds WJ: 1976. Immunologic characterization of canine factor VIII. *Blood* 48: 521–529.

Bouma BN, Dodds WJ, van Mourick JA, Sixma JJ, Webster WP: 1976. Infusion of human and canine factor VIII in dogs with von Willebrand's disease: studies of the von Willebrand and factor VIII synthesis stimulating factors. *Scand J Haematol* 17: 263–275.

Brinkhous KM, Graham JB: 1950. Hemophilia in the female dog. *Science* 111: 723–724.

Brinkhous KM, Griggs TR: 1979. Hemophilia A and B. In: Jones TC, Hackel DB, Migaki G (eds), *Handbook: Animal Models of Human Disease*. Fasc 1 Supplemental Update, 1979, Model No. 12. Washington DC: Registry of Comparative Pathology, Armed Forces Institute of Pathology.

Brinkhous KM, Davis PD, Graham JB, Dodds WJ: 1973. Expression and linkage of genes for X-linked hemophilias A and B in the dog. *Blood* 41: 577–585.

Cotter SM, Brenner RM, Dodds WJ: 1978. Hemophilia A in three unrelated cats. *J Am Vet Med Assoc* 172: 166–168.

Dodds WJ: 1974. Blood coagulation: hemostasis and thrombosis. In: Melby EC Jr, Altman NH (eds), *Handbook of Laboratory Animal Science, II*. Cleveland: CRC Press, pp 85–116.

Dodds WJ: 1981. Second international registry of animal models of thrombosis and hemorrhagic diseases. *ILAR News* 24: R3–R25.

Dodds WJ: 1982. An effective mass-screening program for animal models of inherited bleeding disorders. In: Desnick RJ, Patterson DF, Scarpelli DG (eds), *Animal Models of Inherited Metabolic Diseases*. New York: Alan R. Liss, pp 117–132.

Field RA, Rickard CG, Hutt FB: 1946. Hemophilia in a family of dogs. *Cornell Vet* 36: 285–300.

Graham JB, Buckwalter JA, Hartley LJ, Brinhous KM: 1949. Canine hemophilia: observations on the course, the clotting anomaly, and the effect of blood transfusions. *J Exp Med* 90: 97–111.

Graham JB, Brinkhous KM, Dodds WJ: 1975. Canine and equine hemophilia. In: Brinkhous KM, Hemker HC (eds), *Handbook of Hemophilia, Part 1*. Amsterdam: Excerpta Medica, pp 119–139.

Healy PJ, Sewell CA, Exner T, Morton AG, Adams BS: 1984. Haemophilia in Hereford cattle: factor VIII deficiency. *Aust Vet J* 61: 132 only.

Hutchins DR, Lepherd EE, Crook IG: 1967. A case of equine haemophilia. *Aust Vet J* 43: 83–87.

Hutt FB, Rickard CG, Field RA: 1948. Sex-linked hemophilia in dogs. *J Hered* 39: 3–9.

Mullins LJ, Grant SG, Pazik J, Stephenson DA, Chapman VM: 1987. Assignment and linkage of four new genes to the mouse X chromosome. *Mouse News Lett* 77: 150–151.

Murtaugh RJ, Dodds WJ: 1988. Hemophilia A in a female dog. *J Am Vet Med Assoc* 193: 351–352.

Nossel HL, Archer RK, Macfarlane RG: 1962. Equine haemophilia: report of a case and its response to multiple infusions of heterospecific AHG. *Br J Haematol* 8: 335–342.

Parks BJ, Brinkhous KM, Harris PF, Penick GD: 1964. Laboratory detection of female carriers of canine hemophilia. *Thromb Diath Haemorrh* 12: 368–376.

Sanger VL, Mairs RE, Trapp AL: 1964. Hemophilia in a foal. *J Am Vet Med Assoc* 144: 259–264.

Spurling NW: 1980. Hereditary disorders of haemostasis in dogs: a critical review of the literature. *Vet Bull* 50: 151–173.

*30358 COAGULATION FACTOR IX (HEMOPHILIA B; CHRISTMAS DISEASE; F9) [30690]

Cat

Factor IX deficiency has been described in the British shorthair cat. The disease is a mild spontaneous bleeding diathesis, exacerbated by trauma or surgery. The laboratory findings are identical to those in the human and canine forms of the disease. A colony of these cats is maintained in Albany, New York (Dodds 1981).

Dog

Hemophilia B was described in the dog in 1960 (Mustard et al.1960; Rowsell et al. 1960). It is not as common as hemophilia A (30356) and is known in 11 breeds only. It most frequently occurs as mild to moderate hemorrhaging in smaller breeds and more severe bleeding in larger. The signs are generally similar to those observed in hemophilia A; however, in contrast to the latter, where factor VIII deficiency ranges from severe (<1% activity) to mild (20–30% activity), all reported cases of hemophilia B have had <1% factor IX activity. Carrier females have reduced activity (40–60%) (Dodds and Buckner 1979). Serum and plasma transfusions are equally effective in correcting the clotting defect (Mustard et al. 1962). Brinkhous et al. (1973) bred 2 strains of dogs, each segregating for a different type of hemophilia (see details under 30356). There is considerable evidence that the human and canine forms of the disease are homologous (Brinkhous and Griggs 1979). Two colonies of dogs with this defect are maintained in the United States (Dodds 1981). Spurling (1980) reviewed hemophilia B and other hereditary canine disorders of hemostasis, and Dodds (1982) described a mass-screening program to detect inherited bleeding disorders.

Mouse, *Cf-9*

Hemophilia has not been described in the mouse, but Avner et al. (1987) used an interspecific mouse cross involving *Mus spretus* to show that the human locus for factor IX has an apparently homologous region on the mouse X chromosome; the degree of similarity between the 2 nucleotide sequences appears to be low. Mullins et al. (1987) state that the gene is located in the midpart of the chromosome between *Hprt* and *G6pd*. No mutations at this locus are known.

Virginia opossum (*Didelphis virginiana*)

This species has no highly conserved homolog for human F9 cDNA (Kaslow et al. 1987).

Avner P, Amar L, Arnaud D, Hanauer A, Cambrou J: 1987. Detailed ordering of markers localizing to the Xq26–Xqter region of the human X chromosome by the use of an interspecific *Mus spretus* mouse cross. *Proc Natl Acad Sci USA* 84: 1629–1633.

Brinkhous KM, Griggs TR: 1979. Hemophilia A and B. In: Jones TC, Hackel DB, Migaki G (eds), *Handbook: Animal Models of Human Disease*. Fasc. 1. Supplemental Update, 1979, Model No. 12.Washington DC: Registry of Comparative Pathology, Armed Forces Institute of Pathology.

Brinkhous KM, Davis PD, Graham JB, Dodds WJ: 1973. Expression and linkage of genes for X-linked hemophilia A and B in the dog. *Blood* 41: 577–585.

Dodds WJ: 1981. Second international registry of animal models of thrombosis and hemorrhagic diseases. *ILAR News* 24: R3–R25.

Dodds WJ: 1982. An effective mass-screening program for animal models of inherited bleeding disorders. In: Desnick RJ, Patterson DF, Scarpelli DG (eds), *Animal Models of Inherited Metabolic Disease*. New York: Alan R. Liss, pp 117–132.

Dodds WJ, Buckner RG: 1979. Hemophilia B (factor IX deficiency, Christmas disease). In: Andrews EJ, Ward BC, Altman NH (eds), *Spontaneous Animal Models of Human Disease*. New York: Academic Press, vol 1, pp 271–272.

Kaslow DC, Migeon BR, Persico MG, Zollo M, VandeBerg JL, Samollow PB: 1987. Molecular studies of marsupial X chromosomes reveal limited sequence homology of mammalian X-linked genes. *Genomics* 1: 19–28.

Mullins LJ, Grant SG, Pazik J, Stephenson DA, Chapman VM: 1987. Assignment and linkage of four new genes to the mouse X chromosome. *Mouse News Lett* 77: 150–151.

Mustard JF, Rowsell HC, Robinson GA, Hoeksema TC, Downie HG: 1960. Canine hemophilia B (Christmas disease). *Br J Haematol* 6: 259–266.

Mustard JF, Basser W, Hedgardt G, Secord D, Rowsell HC, Downie HG: 1962. A comparison of the effect of serum and plasma transfusions on the clotting defect in canine haemophilia B. *Br J Haematol* 8: 36–42.

Rowsell HC, Downie HG, Mustard JF, Leeson JE, Archibald JA: 1960. A disorder resembling hemophilia B (Christmas disease) in dogs. *J Am Vet Med Assoc* 137: 247–250.

Spurling NW: 1980. Hereditary disorders of haemostasis in dogs: a critical review of the literature. *Vet Bull* 50: 151–173.

*30403 CONGENITAL TREMOR TYPE AIII [?31160]

Pig, *TRAIII*

There are 2 genetically determined forms of congenital tremor in pigs (synonyms: trembles, trembling, dancing pig disease, myoclonia congenita) characterized by marked deficiency of myelin throughout the CNS, particularly the spinal cord (Done 1976). The X-linked form has been found only in pure Landrace animals or the male offspring of Landrace cross sows (Harding et al. 1973). The disorder is characterized clinically by rhythmic tremors of the limbs and head at rates of 2–8 or more per second, which reduce in amplitude when the animal lies down, and cease when it is asleep (Done 1976). The spinal cord is underdeveloped, and its weight, DNA content, protein content, and total lipid content are markedly decreased (Patterson et al. 1972). These authors could find no evidence of demyelination and claim that their observations are consistent with subnormal myelination and associated with a deficiency of oligodendrocytes. Ultrastructurally, there are many unmyelinated axons; most small or medium-sized axons lack sheaths. Some myelin sheaths are of normal thickness, but most are thinner than normal. There is marked reduction in the number of oligodendrocytes, and there is no evidence of demyelination (Blakemore et al. 1974). Although this mutant shares many features with jimpy in mice, myelin deficiency in the rat, "shaking pups" in the dogs (31208), and the Pelizaeus–Merzbacher disease in man (MIM 31160), Patterson et al. (1972) and Blakemore et al. (1974) point out that there are differences in the pathobiochemistry of congenital tremor and jimpy. To my knowledge there have been no reports of studies on the myelin proteolipid protein in pigs with congenital tremors.

Blakemore WF, Harding JDJ, Done JT: 1974. Ultrastructural observations on the spinal cord of a Landrace pig with congenital tremor type AIII. *Res Vet Sci* 17: 174–178.
Done JT: 1976. Developmental disorders of the nervous system in animals. *Adv Vet Sci Comp Med* 20: 69–114 (p 100).
Harding JDJ, Done JT, Harbourne JF, Randall CJ, Gilbert FR: 1973. Congenital tremor type AIII in pigs: an hereditary sex-linked cerebrospinal hypomyelinogenesis. *Vet Rec* 92: 527–529.
Patterson DSP, Sweasey D, Harding JDJ: 1972. Lipid deficiency in the central nervous system of Landrace piglets affected with congenital tremor AIII, a form of cerebrospinal hypomyelinogenesis. *J Neurochem* 19: 2791–2799.

*30413 CREAM [NK]

Mouse, *Crm*

This mutation occurred in an outbred albino colony (Hetherington 1977). Males and homozygous females have a pale yellow tinge, whereas heterozygotes have yellow patches. The pale yellow phenotype is also expressed on the unpigmented areas of mice with white-spotting (Searle and Beechey 1977). The coat of albino mice carrying the mutation fluoresces with long-wave UV light, and a mosaic pattern of fluorescent and nonfluorescent areas is seen in heterozygotes (Peters and Searle 1980).

Hetherington CM: 1977. Private communication. *Mouse News Lett* 56: 35 only.
Peters J, Searle AG: 1980. Private communication. *Mouse News Lett* 61: 37 only.
Searle AG, Beechey CV: 1977. Private communication. *Mouse News Lett* 57: 18 only.

30415 CUTIS LAXA, X-LINKED (OCCIPITAL HORN TYPE EHLERS–DANLOS SYNDROME; E–D IX)

See discussion under MENKES SYNDROME (30940).

*30432 CYSTINURIA [NK]

Dog

Although cases of cystine stones in dogs had been reported for over a century, cystinuria was first described in a family of pedigreed Irish terriers in the mid-1930s (Morris et al. 1935; Green et al. 1936; Brand et al. 1940). The first 2 animals in which the disorder was observed were males; the second was the son of a male littermate of the original cystinuric animal. Unfortunately, no information was presented about the second dog's dam and her possible relation to the dam of the first dog. No genetic analysis was ever published by the original investigators, but Knox (1966) pieced together information from Brand et al. (1938) and Hess and Sullivan (1942) to create a limited pedigree. The segregation of the disorder was compatible with X-linked inheritance. This colony was dispersed after Brand and Cahill died (Segal and Bovee 1979). A genetic analysis of a cystinuric Scottish terrier kindred and a cystinuric Irish terrier family indicated X-linkage (Tsan et al. 1972a). There has been controversy about the exact amino acid excretion pattern. Crane and Turner (1956) studied a Labrador with cystine stones and found large amounts of cystine and lysine, but normal levels of arginine and ornithine, in the urine. This observation was confirmed by Cornelius et al. (1967) in 4 dachshunds, a basset hound, and a chihuahua, and by Clark and Cuddeford (1971) in 13 dogs of several breeds. Treacher (1962, 1963, 1964a,b, 1966) described abnormal excretion of arginine and ornithine as well as cystine and lysine in dogs of several breeds. Hess and Sullivan (1942) and Treacher (1965) proposed an intestinal defect in amino acid transport. On the basis of his studies, Treacher thought that the canine disease was similar to that in man (MIM 22010). Holtzapple et al. (1969, 1971) and Tsan et al. (1972b) observed normal cystine and lysine accumulation in vitro in intestinal and renal tissue, and suggested that the canine disorder is not similar to that in man. Tsan et al. (1972b) postulated 2 genetically controlled transport systems, which exist in both kidney and intestine. One system, the major pathway transporting cystine, arginine, ornithine, and lysine in the kidneys and intestines, is absent in canine cystinuria. Bovee et al. (1974), on the basis of renal clearance studies, suggested that canine cystinuria may be a metabolic disease manifest as a tubular reabsorptive defect. At present, it seems reasonable, both genetically and metabolically, to suggest that human and canine cystinuria are different disorders.

Other species

Oldfield et al. (1956) reported cystine calculi in a male ranch mink. Bush and Bovee (1978) reported a renal cystine calculus in a male maned wolf (*Chrysocyon brachyurus*). Subsequent studies indicated that the defect is quite common in this species (Bovee et al. 1981). Although the nature of the aminoaciduria appears to be more similar to that found in dogs than that found in humans, the fact that both sexes are affected suggests that the condition is not X-linked. At the autopsy of a male lynx, Jackson and Jones (1979) found 160 microcalculi composed of L-cystine.

Bovee KC, Thier SO, Rea C, Segal S: 1974. Renal clearance of amino acids in canine cystinuria. *Metabolism* 23: 51–58.

Bovee KC, Bush M, Dietz J, Jezyk P, Segal S: 1981. Cystinuria in the maned wolf of South America. *Science* 212: 919–920.

Brand E, Cahill GF, Slanetz CA: 1938. Canine cystinuria. IV. (Abstr) *J Biol Chem* 123: xvi only.

Brand E, Cahill GF, Kassell B: 1940. Canine cystinuria. V. Family history of two cystinuric Irish terriers and cystine determination in dog urine. *J Biol Chem* 133: 431–436.

Bush M, Bovee KC: 1978. Cystinuria in a maned wolf. *J Am Vet Med Assoc* 173: 1159–1162.

Clark WT, Cuddeford D: 1971. A study of the amino-acids in urine from dogs with cystine urolithiasis. *Vet Rec* 88: 414–417.

Cornelius CE, Bishop JA, Schaffer MA: 1967. A quantitative study of aminoaciduria in dachshunds with a history of urolithiasis. *Cornell Vet* 57: 177–183.

Crane CW, Turner AW: 1956. Amino-acid patterns of urine and blood plasma in a cystinuric Labrador dog. *Nature* 177: 237–238.

Green DF, Morris ML, Cahill GF, Brand E: 1936. Canine cystinuria. II. Analysis of cystine calculi and sulfur distribution in the urine. *J Biol Chem* 114: 91–94.

Hess WC, Sullivan MX: 1942. Canine cystinuria. The effect of feeding cystine, cysteine, and methionine at different protein levels. *J Biol Chem* 143: 545–550.

Holtzapple PG, Bovee K, Rea CF, Segal S: 1969. Amino acid uptake by kidney and jejunal tissue from dogs with cystine stones. *Science* 166: 1525–1527.

Holtzapple PG, Rea C, Bovee K, Segal S: 1971. Characteristics of cystine and lysine transport in renal and jejunal tissue from cystinuric dogs. *Metabolism* 20: 1016–1022.

Jackson OF, Jones DM: 1979. Cystine calculi in a caracal lynx (*Felis caracal*). *J Comp Pathol* 89: 39–42.

Know WE: 1966. Cystinuria. In: Stanbury JB, Wyngaarden JB, Frederickson DS (eds), *The Metabolic Basis of Inherited Disease*, 2nd ed. New York: McGraw Hill, pp 1262–1282 (p 1277).

Morris ML, Green DF, Dinkel JW, Brand E: 1935. Canine cystinuria: an unusual case of urinary calculi in the dog. *North Am Vet* 16 (10): 16–20.

Oldfield JE, Allen PH, Adair J: 1956. Identification of cystine calculi in mink. *Proc Soc Exp Biol Med* 91: 560–562.

Segal S, Bovee K: 1979. Canine models of human renal transport disorders. In: Hommes FA (ed), *Models for the Study of Inborn Errors of Metabolism*. Amsterdam: Elsevier/North Holland Biomedical Press, pp 19–31.

Treacher RJ: 1962. Amino-acid excretion in canine cystine-stone disease. *Vet Rec* 74: 503–504.

Treacher RJ: 1963. The amino-aciduria of canine cystine-stone disease. *Res Vet Sci* 4: 556–567.

Treacher RJ: 1964a. The aetiology of canine cystinuria. *Biochem J* 90: 494–498.

Treacher RJ: 1964b. Quantitative studies on the excretion of the basic amino acids in canine cystinuria. *Br Vet J* 120: 178–185.

Treacher RJ: 1965. Intestinal absorption of lysine in cystinuric dogs. *J Comp Pathol Ther* 75: 309–322.

Treacher RJ: 1966. Urolithiasis in the dog. II. Biochemical aspects. *J Small Anim Pract* 1: 537–547.

Tsan MF, Jones TC, Thornton GW, Levy HL, Gilmore C, Wilson TH: 1972a. Canine cystinuria: its urinary amino acid pattern and genetic analysis. *Am J Vet Res* 33: 2455–2461.

Tsan MF, Jones TC, Wilson TH: 1972b. Canine cystinuria: intestinal and renal amino acid transport. *Am J Vet Res* 33: 2463–2468.

*30433 CYTOCHROME b-245, BETA POLYPEPTIDE [30640]

Mouse, *Cybb*

Chapman et al. (1987) and Mullins et al. (1988) referred to this locus in the mouse without giving any details. The locus is in the *Otc–Hprt* region.

Chapman VM, Mullins LJ, Grant SG, Stephenson DA: 1987. X-chromosome gene mapping using two *Mus* species hybrids. *Mouse News Lett* 79: 64 only.

Mullins LJ, Stephenson DA, Grant SG, Chapman VM: 1988. Detailed mapping of the proximal end of the mouse X chromosome. *Mouse News Lett* 80: 181–182.

*30494 DIHYDROTESTOSTERONE RECEPTOR (TESTICULAR FEMINIZATION; ANDROGEN RECEPTOR DEFICIENCY; DHTR) [31370]

Meyer et al. (1975) assigned the locus for the dihydrotestosterone receptor to the human X chromosome and demonstrated that in the mouse a mutation at the locus caused androgen insensitivity. Migeon et al. (1981) used complementation analysis to prove that the locus in the mouse and that in man are homologs.

Chimpanzee

Eil et al. (1980) described testicular feminization in 2 chimpanzees with bilateral inguinal masses that proved to be testes containing Sertoli cell adenomas. Fibroblasts grown from genital skin biopsies yielded no high-affinity, saturable binding for [^{3}H]dihydrotestosterone. The karyotype in both animals was 48,XY.

Cattle

New (1966) described testicular feminization in 3 of 8 single offspring of a cow mated to several bulls. The affected animals had well-developed udders, female external genitalia, and rudimentary male and female ducts. The intraabdominal gonads were testes, and all cells grown from skin cultures had an XY chromosome constitution. The disorder was considered to be either X-linked recessive or sex-influenced autosomal dominant. In view of the existence of a proven X-linked form of the syndrome in other mammals, the former seems more reasonable. I do not know if these mutants are still maintained.

Horse

Individual cases of 64,XY mares with normal external genitalia have been reported (Chandley et al. 1975). Kent et al. (1986) described a syndrome, called XY sex reversal, in which a phenotypic mare is born with the karyotype of a stallion. The syndrome is genetically and clinically heterogeneous. One form appears to be X-linked, although autosomal sex-limited transmission cannot be ruled out. The phenotypic expression ranges from a feminine mare with a reproductive tract within normal limits to a greatly masculinized mare. Among 27 horses whose gonads were described by Kent et al., 8 had "ovatestes ranging to testicular feminization." See 31491.

Mouse, testicular feminization, *Tfm*

Fekete (1938) reported what appears to have been testicular feminization in 10 closely related animals of the Little–Murray dilute brown strain. All 10 were normal females externally. Internally, a short vagina ended blindly, and there were no female sex organs; the gonads were testes that varied in size and development. Apparently this mutation could not be maintained. Thirty-two years later, Lyon and Hawkes (1970) described the effects of a recurrence of this mutation and determined its linkage. From birth until the time of normal sexual maturity, the feminized males are indistinguishable externally from normal females. In some affected animals, the vagina opens much later than it does in normal females. The feminized males behave like nonestrus females. Internally, there are very small testes. The vas deferans and epididymis are absent, and there are no male accessory glands. Aside from a small vagina, there are no female reproductive organs. During embryogenesis, the vagina forms as in the female but subsequently degenerates, probably as a result of the action of the anti-Müllerian hormone. At the end of the degeneration process, a small vaginal pocket, which varies in the adult Tfm mouse, remains (Mauch et al. 1985). In the testes, spermatogonia (some in mitosis) and Sertoli cells are present, but spermatogenesis ends at the spermatocyte stage or earlier; the tubules are hyalinized, and, in older animals, the Leydig cells are hypertrophied. The seminiferous tubule barrier in *Tfm*/Y animals is impaired (Fritz et al. 1983). Affected males have a 40, XY karyotype and express the H–Y antigen (Bennett et al. 1975). Ohno and Lyon (1970) demonstrated that kidney alcohol dehydrogenase is not inducible by testosterone administration, and proposed that *Tfm* is a noninducible mutation of a repressive regulatory locus that mediates the manifestation of the male phenotype. Gehring et al. (1971) and Bullock et al. (1971a) reported that defective nuclear retention of androgen occurs in the kidney of the Tfm mouse, and suggested that the cytosol and nuclear 5alpha-dihydrotestosterone receptor protein might be a regulatory protein specified by the locus. Bullock and Bardin (1972) showed that the cytosol from the kidney of affected animals has no demonstrable specific androgen receptor, and postulated that this deficiency precludes androgen transfer to the nucleus of the cell, resulting in androgen insensitivity. Further evidence that the androgen insensitivity is due to a deficiency of a cystol androgen-receptor complex has come from the studies of Bardin et al. (1973), Bullock and Bardin (1974, 1975), Attardi and Ohno (1974), Gehring and Tomkins (1974), and Verhoeven and Wilson (1976). This receptor complex in the kidney also binds estrogens (Bullock and Bardin 1975). In addition, the kidney possesses a distinct high-affinity estrogen-receptor complex that is not affected by the *Tfm* mutation. The androgen receptor in heterozygotes has the same physical properties as that of normal mice, but, owing to a decrease in concentration, binding activity is approximately 2/3 of normal (Bullock et al. 1975). Bulfield and Nahum (1978) showed that the *Tfm* mutation is pleiotropic and prevents repression of the androgen-repressible enzyme ornithine aminotransferase. Thus, both inducible and repressible systems share the same cystol androgen-receptor. The same authors demonstrated that estrogen-inducible enzymes are not affected by the mutation. While investigating the pathogenesis of the disorder, Goldstein and Wilson (1972) administered dihydrotestosterone to pregnant carriers and studied the effect on their male, female, and *Tfm*/Y offspring. Normal female offspring demonstrated striking virilization of the internal genital tract, but no virilization occurred in *Tfm*/Y animals. This observation provides direct evidence that the disorder results from resistance to androgen action during androgen-mediated sexual differentiation during embryogenesis. The submandibular (submaxillary) salivary gland normally exhibits sexual dimorphism, and in Tfm mice it histologically resembles that observed in females and castrated males (Andrews and Bullock 1972; Lyon et al. 1973); it is insensitive to androgens (Lyon et al. 1973), and lacks an esteroproteolytic enzyme (Schenkein et al. 1974) and an androgen-binding complex present in normal mice (Verhoeven and Wilson 1976). Brain cytosol from *Tfm*/Y mice contains some androgen receptors (20–25% of normal); these are similar to those of normal mice but less aggregated (Attardi et al. 1976; Attardi and Ohno 1978). Wieland et al. (1978), Wieland and Fox (1979), and Fox et al. (1983) have further characterized these residual androgen receptors. The results of the last-mentioned study indicate that the residual androgen receptors, in kidney as well as brain, appear to be like those of a normal minor receptor species; there is no evidence of the wild-type species. In contrast, the residual receptors in animals carrying a second *Tfm* mutation reported by Lovell (see below), although less prevalent than in animals carrying the original mutation, appear to be like the wild-type receptors. In this respect, they are like the receptors observed in Tfm rats (see below). These observations suggest that the 2 mutations in the mouse are alleles, not occurrences of the same mutation, and that the one reported by Lovell may be identical with *Tfm* in the rat. Tettenborn et al. (1971) found that in 3 of 58 heterozygotes the response of kidney enzymes to administered testosterone was indistinguishable from that of affected animals, and, in an additional 5 heterozygotes, the response was intermediate between +/+ females and *Tfm*/Y males. Heterozygotes responded to testosterone in the submaxillary test (Lyon et al. 1973), but the response appeared to be intermediate in some. In normal male and female mice, testosterone exercises a negative feedback upon pituitary gonadotrophs, which hypertrophy after castration or spaying and atrophy in response to prolonged administration of testosterone. However, the gonadotrophs of Tfm animals atrophy after castration and show marked hypertrophy in response to testosterone (Itakura and Ohno 1973). A closely linked "controlling element," probably the chromosomal controlling element (*Xce*) (31467), can modify the expression of the *Tfm* gene to the extent that *Tfm*/Y animals develop a number of male characteristics (Ohno et al. 1973). However, Blecher (1978) claims that Tfm mice possess rete testis, efferent ductules, and microscopic epididymes bilaterally, and says there is no need to evoke the operation of a "controlling element." His opinion is based on a study of 2 animals said to be free of the controlling factor, but no evidence to support this is given. The microscopic epididymis described by Blecher possessed beta-glucuronidase activity (Scott and Blecher 1987). Approximately 20% of affected animals over 600 days old develop testicular germ cell tumors (Ohno 1974); Chung (1980) described ultrastructural studies of these tumors. Using chimeras of genotype *Tfm*/Y ↔ +/Y, Lyon et al. (1975) provided evidence that male germ cells possessing the *Tfm* gene are capable of normal maturation and fertilization. Ohno (1976, 1977, 1979) stated that the *Tfm* locus specifies the nuclear–cystol androgen-receptor protein, and that it and the H–Y antigen are the 2 major regulatory genes for mammalian sex determination. *Tfm*/*Tfm* females, created by using males chimeric for the *Tfm* gene, are fertile; however, their reproductive performance is impaired and stops prematurely (Lyon and Glenister 1980). *Tfm*/+ females reproduce normally. Takeda et al. (1987) used steroid autoradiographic techniques to demonstrate mosaicism of X-chromosome inactivation in 2 androgen target tissues – mesenchyme of the urogenital sinus and mammary gland rudiments – of heterozygote fetuses.

Tfm^{Lac}

Lovell (1978) described a second mutation that initiates testicular feminization. The phenotype of affected males is virtually identical with that of *Tfm*/Y animals, except that the testes appear to be slightly larger; measurements of wet weight, protein, and DNA content indicate they are significantly larger (Fox et al. 1983). Studies on the residual androgen receptors suggest that the 2 mutations are alleles and not occurrences of the same mutation (Fox et al. 1983, and discussion above). *Note:* The use of *Tfm*^{H} and *Tfm*^{L} to designate the 2 alleles (Fox et al. 1983) is unconventional.

Pig

Lojda (1975) described a syndrome similar to testicular feminization among offspring of a Large White × Landrace sow mated to 4 unrelated boars. The disorder was transmitted to a subsequent generation through a normal female offspring. Affected animals had normal female external genitalia, a blind and frequently hypoplastic vagina, and testes. There are a number of disturbing aspects about this study, including a significantly distorted primary sex ratio in favor of males and a high frequency of 38 XX/XY mosaicism among the affected piglets.

Rat, testicular feminization, *Tfm*

D'Amour and Funk (1941) described the occurrence of 8 male pseudohermaphrodites in a large colony of rats. No genetic or endocrinologic studies were done, but, on the basis of anatomical description, these animals probably had a form of testicular feminization. This mutant is extinct (Stanley et al. 1973). Another independent occurrence of the defect was described and designated pseudohermaphroditism (*Ps*) by Allison et al. (1965) and Bardin et al. (1969, 1970). The phenotype is that of a female with a short vagina, small phallus, and a well-developed nipple line. The gonads are testes located in the inguinal canal; no other male reproductive organs develop. Müllerian and Wolffian duct systems develop normally up to day 17 of gestation, at which time they begin to disintegrate; the process is complete by day 19 (Stanley et al. 1973). Sperm cells fail to develop beyond primary spermatocytes, and there is Leydig cell hyperplasia, which is sensitive to BCG stimulation (Vanha-Perttula et al. 1970); Chung and Hamilton (1976) described the fine structure of the Leydig cells. In a large proportion of affected animals aged 1–3 years, Sertoli cell tumors develop in one and occasionally both testes (Stanley et al. 1973); these tumors have been described in more detail by Allison et al. (1978) and Chung et al. (1980). The pattern of transmission is consistent with X-linked or sex-linked dominant inheritance (Stanley et al. 1973); in view of the proven X-linkage of the apparently homologous condition in the other mammals, the former seems more reasonable. End organ insensitivity to testosterone administration is generalized (Allison et al. 1971; Bullock et al. 1971a,b; Grossman et al. 1971; Neuhaus and Irwin 1972); however, this appears to be relative rather than absolute (Sherins and Bardin 1971; Sherins et al. 1971); and in this regard the syndrome in the rat is not as extreme as in the mouse. Affected animals are unable to concentrate dihydrotestosterone in the cells (Bardin et al. 1973). This cannot be explained by a decrease in total activity of 5alpha-steroid reductase in target tissues (Bullock et al. 1970), and reduction of testosterone to dihydrotestosterone is normal (Bullock and Bardin 1973). The latter investigators concluded that the androgen insensitivity is the result of an inability to concentrate dihydrotestosterone at its site of action. (Incidentally, in this report the symbol *Tfm* replaced *Ps*.) Ritzen et al. (1972) described deficient nuclear uptake of testosterone in affected animals. Bullock and Bardin (1972) showed that the cytosol from the kidney of affected animals had no demonstrable specific androgen receptor and postulated that this deficiency precludes androgen transfer to the nucleus. Bardin et al. (1973) provided further evidence for this position. Although the androgen receptors in the anterior pituitary in the brain of *Tfm*/Y rats are severely reduced in numbers (10–15% of normal), their physicochemical properties are indistinguishable from those of normal rats (Naess et al. 1976); Wieland and Fox (1981) and Fox et al. (1983) further characterized the residual androgen receptors and concluded that the situation in the *Tfm*/Y is similar to that in the *Tfm^{Lac}*/Y mouse (see above) and certain patients with TFM; the residual androgen receptors are of the same type as those found in normal males. A cytosolic androgen receptor is present in skeletal muscle from *Tfm*/Y rats, but it does not bind DNA (Max 1981).

Allison JE, Stanley AJ, Gumbreck LG: 1965. Sex chromatin and idiograms from rats exhibiting anomalies of the reproductive organs. *Anat Rec* 153: 85–92.

Allison JE, Chan F, Stanley AJ, Gumbreck LG: 1971. Androgen insensitivity in male pseudohermaphrodite rats. *Endocrinology* 89: 615–617.

Allison JE, Weidenbach PW, Becker RR, Hollander R: 1978. Hormonal factors related to the incidence of testicular tumors in male pseudohermaphrodite rats. *Fertil Steril* 30: 91–95.

Andrews EJ, Bullock LP: 1972. A morphological and histochemical evaluation of sexual dimorphism in androgen-insensitive pseudohermaphroditic mice. *Anat Rec* 174: 361–370.

Attardi B, Ohno S: 1974. Cytosol androgen receptor from kidney of normal and testicular feminized (*Tfm*) mice. *Cell* 2: 205–212.

Attardi B, Ohno S: 1978. Physical properties of androgen receptors in brain cytosol from normal and testicular feminized (*Tfm*/y) mice. *Endocrinology* 103: 760–770.

Attardi B, Geller LN, Ohno S: 1976. Androgen and estrogen receptors in brain cytosol from male, female, and testicular feminized (tfm/y) mice. *Endocrinology* 98: 864–874.

Bardin CW, Allison JE, Stanley AJ, Gumbreck LG: 1969. Secretion of testosterone by the pseudohermaphrodite rat. *Endocrinology* 84: 435–436.

Bardin CW, Bullock L, Schneider G, Allison JE, Stanley AJ: 1970. Pseudohermaphrodite rat: end organ insensitivity to testosterone. *Science* 167: 1136–1137.

Bardin CW, Bullock LP, Sherins RJ, Mowszowicz I, Blackburn WR: 1973. Part II. Androgen metabolism and mechanism of action in male pseudohermaphroditism: a study of testicular feminization. *Recent Prog Horm Res* 29: 65–109.

Bennett D, Boyse EA, Lyon MF, Mathieson BJ, Scheid M, Yanagisawa K: 1975. Expression of H–Y (male) antigen in phenotypically female *Tfm*/Y mice. *Nature* 257: 236–238.

Blecher SR: 1978. Microscopic epididymides in testicular feminisation. *Nature* 275: 748–749.

Bulfield G, Nahum A: 1978. Effect of mouse mutants testicular feminization and sex reversal on hormone-mediated induction and repression of enzymes. *Biochem Genet* 16: 743–750.

Bullock LP, Bardin CW: 1972. Androgen receptors in testicular feminization. *J Clin Endocrinol Metab* 35: 935–937.

Bullock LP, Bardin CW: 1973. *In vivo* and *in vitro* testosterone metabolism by the androgen insensitive rat. *J Steroid Biochem* 4: 139–151.

Bullock LP, Bardin CW: 1974. Androgen receptors in mouse kidney: a study of male and female and androgen-insensitive (tfm/y) mice. *Endrocrinology* 94: 746–756.

Bullock LP, Bardin CW: 1975. The presence of estrogen receptor in kidneys from normal and androgen-insensitive tfm/y mice. *Endocrinology* 97: 1106–1111.

Bullock LP, Schneider G, Bardin CW: 1970. 5Alpha-steroid reductase activity in tissues from rats with end organ insensitivity to testosterone. *Life Sci* 9(I): 701–705.

Bullock LP, Bardin CW, Ohno S: 1971a. The androgen insensitive mouse: absence of intranuclear androgen retention in the kidney. *Biochem Biophys Res Commun* 44: 1537–1543.

Bullock LP, Bardin CW, Gram TE, Schroeder DH, Gillette JR: 1971b. Hepatic ethylmorphine demethylase and delta^4-steroid reductase in the androgen-insensitive pseudoherma-phroditic rat. *Endocrinology* 88: 1521–1523.

Bullock LP, Mainwaring WIP, Bardin CW: 1975. The physico-chemical properties of the cytoplasmic androgen receptor in the kidneys of normal, carrier female (tfm/+) and androgen-insensitive (tfm/y) mice. *Endocr Res Commun* 2: 25–45.

Chandley AC, Fletcher J, Rossdale PD, Peace CK, Ricketts SW, McEnery RJ, Thorne JP, Short RV, Allen WR: 1975. Chromosome abnormalities as a cause of infertility in mares. *J Reprod Fertil* (Suppl) 23: 377–383.

Chung KW: 1980. Spontaneous testicular neoplasm in mice with testicular feminization. *Cell Tissue Res* 208: 47–56.

Chung KW, Hamilton JB: 1976. Further observations on the fine structure of Leydig cells in the testes of male pseudohermaphrodite rats. *J Ultrastruct Res* 54: 68–75.

Chung KW, Allison JE, Stanley AJ: 1980. Structural and functional factors related to testicular neoplasia in feminized rats. *J Natl Cancer Inst* 65: 161–165.

D'Amour FE, Funk D: 1941. Sponataneous intersexuality in the rat. *Endocrinology* 28: 727–728.

Eil C, Merriam GR, Bowen J, Ebert J, Tabor E, White B, Douglass EC, Loriaux DL: 1980. Testicular feminization in the chimpanzee. (Abstr) *Clin Res* 28: 624A only.

Fekete E: 1938. Sexual abnormalities in an inbred strain of mice. *Proc Soc Exp Biol Med* 38: 59–62.

Fox TO, Blank D, Politch JA: 1983. Residual androgen binding in testicular feminization (TFM). *J Steroid Biochem* 19: 577–581.

Fritz IB, Lyon MF, Setchell BP: 1983. Evidence for a defective seminiferous tubule barrier in testes of *Tfm* and *Sxr* mice. *J Reprod Fertil* 67: 359–363.

Gehring U, Tomkins GM: 1974. Characterization of a hormone receptor defect in the androgen-insensitivity mutant. *Cell* 3: 59–64.

Gehring U, Tomkins GM, Ohno S: 1971. Effect of the androgen-insensitivity mutation on a cytoplasmic receptor for dihydrotestosterone. *Nature [New Biol]* 232: 106–107.

Goldstein JL, Wilson JD: 1972. Studies on the pathogenesis of the pseudohermaphroditism in the mouse with testicular feminization. *J Clin Invest* 51: 1647–1658.

Grossman SH, Axelrod B, Bardin CW: 1971. Effect of testosterone on renal and hepatic L-gulonolactonase activities in male, female and pseudohermaphroditic rats. *Life Sci* 10 (II): 175–180.

Itakura H, Ohno S: 1973. The effect of the mouse X-linked *testicular feminization* mutation on the hypothalamus–pituitary axis. 1. Paradoxical effect of testosterone upon pituitary gonadotrophs. *Clin Genet* 4: 91–97.

Kent MG, Shoffner RN, Buoen L, Weber AF: 1986. XY sex-reversal syndrome in the domestic horse. *Cytogenet Cell Genet* 42: 8–18.

Lojda L: 1975. The cytogenetic pattern in pigs with hereditary intersexuality similar to the syndrome of testicular feminization in man. *Docum Vet (Brno)* 8: 71–82.

Lovell D: 1978. Private communication. *Mouse News Lett* 58: 40 only.

Lyon MF, Glenister PH: 1980. Reduced reproductive performance in androgen-resistant *Tfm/Tfm* female mice. *Proc R Soc Lond [Biol]* 208: 1–12.

Lyon MF, Hawkes SG: 1970. X-linked gene for testicular feminization in the mouse. *Nature* 227: 1217–1219.

Lyon MF, Hendry I, Short RV: 1973. The submaxillary salivary glands as test organs for response to androgen in mice with testicular feminization. *J Endocrinol* 58: 357–362.

Lyon MF, Glenister PH, Lamoreux ML: 1975. Normal spermatozoa from androgen-resistant germ cells of chimaeric mice and the role of androgen in spermatogenesis. *Nature* 258: 620–622.

Mauch RB, Thiedemann K-U, Drews U: 1985. The vagina is formed by downgrowth of Wolffian and Müllerian ducts: graphical reconstructions from normal and Tfm mouse embryos. *Anat Embryol* 172: 75–87.

Max SR, 1981. Cytosolic androgen receptor in skeletal muscle from normal and testicular feminization mutant (Tfm) rats. *Biochem Biophys Res Commun* 101: 792–799.

Meyer WJ III, Migeon BR, Migeon CJ: 1975. Locus on human X chromosome for dihydrotestosterone receptor and androgen insensitivity. *Proc Natl Acad Sci USA* 72: 1469–1472.

Migeon BR, Brown TR, Axelman J, Migeon CJ: 1981. Studies of the locus for androgen receptor: localization on the human X chromosome and evidence for homology with the *Tfm* locus in the mouse. *Proc Natl Acad Sci USA* 78: 6339–6343.

Naess O, Haug E, Attramadal A, Aakvaag A, Hansson V, French F: 1976. Androgen receptors in the anterior pituitary and central nervous system of the androgen "insensitive" (Tfm) rat: correlation between receptor binding and effects of androgens on gonadotropin secretion. *Endocrinology* 99: 1295–1303.

Nes N: 1966. Testikulaer feminisering hos storfe. (Norwegian) *Nord Vet-Med* 18: 19–29.

Neuhaus OW, Irwin JF: 1972. Non-inducibility of the sex-dependent urinary protein in male pseudohermaphroditic rats. *Life Sciences* 11 (II): 631–636.

Ohno S: 1974. Animal model: X-linked testicular feminization mutation of the mouse. *Am J Pathol* 76: 589–592.

Ohno S: 1976. Major regulatory genes for mammalian sexual development. *Cell* 7: 315–321.

Ohno S: 1977. The Y-linked H–Y antigen locus and the X-linked *Tfm* locus as major regulatory genes of the mammalian sex determining mechanism. *J Steroid Biochem* 8: 585–592.

Ohno S: 1979. *Major Sex-Determining Genes.* Berlin: Springer-Verlag, pp 104–128.

Ohno S, Lyon MF: 1970. X-linked testicular feminization in the mouse as a non-inducible regulatory mutation of the Jacob–Monod type. *Clin Genet* 1: 121–127.

Ohno S, Christian L, Attardi BJ, Kan J: 1973. Modification of expression of the *testicular feminization (Tfm)* gene of the mouse by a "controlling element" gene. *Nature [New Biol]* 245: 92–93.

Ritzen EM, Nayfeh SN, French FS, Aronin PA: 1972. Deficient nuclear uptake of testosterone in the androgen-insensitive (Stanley–Gumbreck) pseudohermaphrodite male rat. *Endocrinology* 91: 116–124.

Schenkein I, Levy M, Bueker ED, Wilson JD: 1974. Immunological and enzymatic evidence for the absence of an esteroproteolytic enzyme (protease "D") in the submandibular gland of the Tfm mouse. *Endocrinology* 94: 840–844.

Scott JE, Blecher SR: 1987. Beta-glucuronidase activity is present in the microscopic epididymis of the *Tfm*/Y mouse. *Dev Genet* 8: 11–15.

Sherins RJ, Bardin CW: 1971. Preputial gland growth and protein synthesis in the androgen-insensitive male pseudohermaphroditic rat. *Endocrinology* 89: 835–841.

Sherins RJ, Bullock L, Gay VL, Vanha-Perttula T, Bardin CW: 1971. Plasma LH and FSH levels in the androgen insensitive pseudohermaphroditic rat: responses to steroid administration. *Endocrinology* 88: 763–770.

Stanley AJ, Gumbreck LG, Allison JE, Easley RB: 1973. Part I. Male pseudohermaphroditism in the laboratory Norway rat. *Recent Prog Horm Res* 29: 43–64.

Takeda H, Suzuki M, Lasnitzki I, Mizuno T: 1987. Visualization of X-chromosome inactivation mosaicism of *Tfm* gene in X^{Tfm}/X^+ heterozygous female mice. *J Endocrinol* 114: 125–129.

Tettenborn U, Dofuku R, Ohno S: 1971. Noninducible phenotype exhibited by a proportion of female mice heterozygous for the X-linked testicular feminization mutation. *Nature [New Biol]* 234: 37–40.

Vanha-Perttula T, Bardin CW, Allison JE, Gumbreck LG, Stanley AJ: 1970. "Testicular feminization" in the rat: morphology of the testis. *Endocrinology* 89: 611–619.

Verhoeven G, Wilson JD: 1976. Cytosol androgen binding in submandibular gland and kidney of the normal mouse and the mouse with testicular feminization. *Endocrinology* 99: 79–92.

Wieland SJ, Fox TO: 1979. Putative androgen receptors distinguished in wild-type and testicular-feminized (*Tfm*) mice. *Cell* 17: 781–787.

Wieland SJ, Fox TO: 1981. Androgen receptors from rat kidney and brain: DNA-binding properties of wild-type and *tfm* mutant. *J Steroid Biochem* 14: 409–414.

Wieland SJ, Fox TO, Savakis C: 1978. DNA-binding of androgen and estrogen receptors from mouse brain: behavior of residual androgen receptor from *Tfm* mutant. *Brain Res* 140: 159–164.

*30510 ECTODERMAL DYSPLASIA, ANHIDROTIC (HYPOHIDROTIC) (EDA)

Cattle

Drieux et al. (1950) reported congenital hypotrichosis with anadontia in 3 male calves that were completely hairless at birth although they later acquired a sparse coat of abnormal hair. One animal began to grow fine, soft hair at 15 days, but the other 2 were hairless at 5 and 6 weeks. One of these was kept alive for 6 months; a few sparsely scattered hairs developed on the neck at 2 1/2 months, and similar hairs were present on other parts of the body by 4 months. The calves were toothless at birth; the animal kept till 6 months developed 2 upper teeth placed symmetrically, in positions corresponding to the first molars. In all 3 calves, the tongue was larger than normal and, although they were hornless, horn development was retarded but not suppressed; the hooves were normal. Microscopic examination of the skin revealed that the sudoriparous glands were cystic and lacked secretory tubules. X-linkage was concluded on the basis of the sex of the calves, the fact that the mother of 1 calf was the daughter of the cow that produced the other 2, and the fact that, although the calves shared a common sire, he was of a breed quite different from that of the mother and had sired over 180 normal calves by other cows in the same district. The mother of 1 of the affected calves had produced 2 normal male calves by the same bull. The heterozygous females were reported to be normal. Wijeratne et al. (1988) reported a similar defect in male offspring of 2 Friesian cows; however, the calves had normal hair at birth, and there was no macroglossia. The 2 cows, half-sibs by a common dam, produced 4 affected bull calves and 3 normal heifers by 6 unrelated Holstein, Friesian, Devon, and Hereford bulls. The 2 cows and their dam had short, stubby, lusterless hair; pigmented areas were rusty gray rather than black. Two of the affected calves were not examined in detail and died within 2 weeks. In the other 2, the skin of the face, neck, ears, thorax, spinal region, tail, and the inner side of the thighs had a thin coat of fine, short, silky hairs. The normal hair of the abdomen was clearly demarcated from the thin hair of the thorax. Some normal hair was also present on the ventral neck, axillae, groin, and medial aspect of the pinnae. The eyelashes, vibrissae, and tail brush were present. In skin samples taken from multiple sites, the number of large, first-formed hair follicles was reduced. Small-caliber follicles were present, but, unlike those of normal newborn calves, they were in the inactive phase. Sweat glands, sebaceous glands, and arrector pili muscles were normal. The 2 calves lacked incisors. The correlation of this disorder, X-linked dominant congenital hypotrichosis and incisor anodontia, to the one described by Drieux et al. (1950) and to human anhidrotic ectodermal dysplasia remains to be determined.

Dog, congenital ectodermal defect

Selmanowitz et al. (1970) described 2 male miniature French poodles with absent cutaneous appendages (hair follicles, arrector pili muscles, sebaceous glands, and sweat glands). The animals were from the same litter and were the only males; 3 female littermates were normal. The defect was symmetric and the haircoat was absent from about 2/3 of the integument including the head, ventral trunk, dorsal pelvic region, and proximal portion of the limbs. The nails and teeth were normal. In a subsequent brief report in 1979, Selmanowitz added the following to the original description: "Dental anomalies may be found"; the pigment in the remaining haircoat of "occasional affected and nonviable offspring" is diluted; and although the pattern of hair loss is remarkably similar in affected animals, some variation in the extent of involvement exists. A male whippet with a strikingly similar defect was described by Thomsett (1961). Although Selmanowitz et al. (1969) suggest the possibility of X-linkage, and Selmanowitz (1979) states that the disorder is "sex-linked," no genetic data to support this position have been published to my knowledge.

Mouse, tabby, *Ta*

This mutation, which arose spontaneously, was one of the first proven cases of X-linkage in the mouse (Falconer 1952, 1953). Hemizygous males and homozygous females are characterized by the presence of a single, aberrant hair type, abnormal awl, that occurs in place of the guard hairs, zigzags, auchenes, and awls normally present; a bald patch behind each ear; a virtually bald tail with a few kinks near the tip; abnormal dentition; reduced aperture of the eyelids; absence of the tarsal (Meibomian) gland; complete absence or reduction of many, but not all, exocrine glands; absence of the plicae digitales of the ventral surface of the paws; an anomaly of the papillae valletae of the tongue; thin skin; and a respiratory disorder. The tail kinks and the respiratory disorder appear to be secondary to the skin defect and defective nasalglands, respectively. Tabby males breed well, but homozygous females are often sterile. Heterozygous females have transverse stripes reminiscent of the tabby markings of the cat, hence the original assignment of the name. The dark stripes have no agouti bands on the hairs. Guard hairs and zigzags are present with a low frequency. Heterozygotes are fertile. Detailed analyses of the effect of the gene on coat structure, follicles, vibrissae, and tail in affected animals and heterozygotes were carried out by Dun (1959), Grüneberg (1966b, 1969, 1971b), Kindred (1967), and Claxton (1967). The formation of new hair follicles is suppressed prenatally, hair caliber is reduced, and the coat is not differentiated into hair types. Grüneberg (1965, 1966a), Sofaer (1969a), and Miller (1978) have described in detail characteristic abnormalities of the incisors and molars. The former may be reduced or absent, the latter are reduced in size, and many third molars are missing. A phenomenon known as "twinning," in which an additional molar occurs, has been described by Sofaer (1969a); its expression is enhanced by strain A background. Some $Ta/+$ females have a mixture of normal molars, tabby molars, and molars that combine features of both. In some heterozygotes, the "twinning" may occur rarely (Grüneberg 1966a). Many, but not all, exocrine glands are absent or reduced (Grüneberg 1971a,b). Among those considered to be unaffected were the large salivary glands, including the submandibular gland. However, more detailed studies by Blecher et al. (1983) have revealed that the adult gland is reduced in mass and that its granular convoluted tubules are delayed developmentally and reduced in content; these observations suggest that ectodermal growth factor, produced by the tubules, may be deficient in tabby mice. The "tabby syndrome," which is also present in crinkled (cr) and downless (dl) mice, appears to result from interference with the downgrowth of solid epithelial buds into the underlying mesenchyme beginning at about 12 1/2 – 13 days of embryogenesis, when this type of solid bud begins to be formed. Grüneberg favored a basic error in the epithelium rather than in the underlying mesenchyme. Sofaer's culture experiments using tail skin provided no evidence of a primary epidermal effect (Sofaer 1974). However, Mayer and Green (1978) clearly demonstrated, by the use of dermal–epidermal recombination graftings of body skin, that the tabby locus acts within the epidermis and has no effect on the dermis. Pennycuik and Raphael (1984) showed that this difference could be explained by a difference in the site of action of follicle initiation, examined by Mayer and Green. Pennycuik and Raphael suggest that the Ta locus produces a substance that diffuses out of the cell and affects the matrix separating the dermis and epidermis; the matrix, in turn, affects the transmission of signals from the former to the latter. In his extensive writings on this mutant, Grüneberg was concerned with determining whether the phenotypic expression of the gene in $Ta/+$ females could be explained by the X-chromosome inactivation hypothesis of Lyon (1963, 1970) or by a threshold effect; he believed that the evidence overwhelmingly favored the latter. Attempts at resolution using experimentally produced chimeras have given equivocal results (Cattanach et al. 1972; Grüneberg et al. 1972; McLaren et al. 1973). The Ta locus lies centrally in the X chromosome and is closely linked to Xce, Phk, sla, Tfm, and Gs, which lie proximally, and Bhd, Slf, $Pgk-1$, Mo, and Ym, which lie distally (Roderick and Davisson 1986). Lyon (1974) and McKusick (1986) suggest that tabby is homologous with the X-linked form of anhidrotic ectodermal dysplasia characterized by absence of teeth, hypotrichosis, and absence of sweat glands. Although the agreement does not appear to be precise because tabby mice have teeth, Blecher et al. (1982) indicate that the dental phenotypes of the 2 disorders are similar and suggest that they are true homologs.

Ta^J

Stevens (1963) described an allele that resembles *Ta* except that the affected animals have some hair on the tails and the tails are curved. *Ta/Ta^J* females have both hairless areas and areas with curved hair on their tails.

Ta^c

This is another allele with identical expression to *Ta^J* except that the hairs on the tail are not curved (Roberts 1966). The tail hairs are abnormal and are sparser than in the normal mouse (Sofaer 1969b). According to Sofaer (1979), this allele is a recurrence of *Ta^J*.

Blecher SR, Arnold K, Weeks N, Debertin M, Kroone R, Heller N-H: 1982. Possible homolog of ectodermal dysplasia and tabby, and possible role of EGF. (Abstr) *Anat Rec* 202: 17A only.

Blecher SR, Debertin M, Murphy JS: 1983. Pleiotropic effect of tabby gene on epidermal growth factor containing cells of mouse submandibular gland. *Anat Rec* 207: 25–29.

Cattanach BM, Wolfe HG, Lyon MF: 1972. A comparative study of the coats of chimaeric mice and those of heterozygotes for X-linked genes. *Genet Res* 19: 213–228.

Claxton JH: 1967. The initiation and development of hair follicle population in tabby mice. *Genet Res* 10: 161–171.

Drieux H, Priouzeau M, Thiéry G, Priouzeau M-L: 1950. Hypotrichose congénitale avec anodontie, acérie et macroglossie chez le Veau. *Rec Med Vet* 126: 385–399.

Dun RB: 1969. The development and growth of vibrissae in the house mouse with particular reference to the time of action in the tabby (*Ta*) and ragged (*Ra*) genes. *Aust J Biol Sci* 12: 312–330.

Falconer DS: 1952. A totally sex-linked gene in the house mouse. *Nature* 169: 664–665.

Falconer DS: 1953. Total sex-linkage in the house mouse. *Z Indukt Abstammungs-Vererbungsl* 85: 210–219.

Grüneberg H: 1965. Genes and genotypes affecting the teeth of the mouse. *J Embryol Exp Morphol* 14: 137–159.

Grüneberg H: 1966a. The molars of the tabby mouse, and a test of the "single-active X-chromosome" hypothesis. *J Embryol Exp Morphol* 15: 223–244.

Grüneberg H: 1966b. More about the tabby mouse and the Lyon hypothesis. *J Embryol Exp Morphol* 16: 569–590.

Grüneberg H: 1969. Threshold phenomena versus cell heredity in the manifestation of sex-linked genes in mammals. *J Embryol Exp Morphol* 22: 145–179.

Grüneberg H: 1971a. The glandular aspects of the tabby syndrome in the mouse. *J Embryol Exp Morphol* 25: 1–19.

Grüneberg H: 1971b. The tabby syndrome in the mouse. *Proc R Soc Lond [Biol]* 179: 139–156.

Grüneberg H, Cattanach BM, McLaren A, Wolfe HG, Bowman P: 1972. The molars of tabby chimaeras in the mouse. *Proc R Soc Lond [Biol]* 182: 183–192.

Kindred B: 1967. Some observations on the skin and hair of tabby mice. *J Hered* 58: 197–199.

Lyon M: 1963. Attempts to test the inactive-X theory of dosage compensation in mammals. *Genet Res* 4: 93–103.

Lyon MF: 1970. Genetic activity of sex chromosomes in somatic cells of mammals. *Philos Trans R Soc Lond [Biol]* 259: 41–52.

Lyon MF: 1974. Mechanisms and evolutionary origins of variable X-chromosome activity in mammals. *Proc R Soc Lond [Biol]* 187: 243–268.

Mayer TC, Green MC: 1978. Epidermis is the site of action of tabby (*Ta*) in the mouse. *Genetics* 90: 125–131.

McKusick VA: 1986. *Mendelian Inheritance in Man: Catalogs of Autosomal Dominant, Autosomal Recessive, and X-Linked Phenotypes*, 7th ed. Baltimore: Johns Hopkins Univ Press.

McLaren A, Gauld IK, Bowman P: 1973. Comparison between mice chimaeric and heterozygous for the X-linked gene *tabby*. *Nature* 241: 180–183.

Miller WA: 1978. The dentitions of tabby and crinkled mice (an upset in mesodermal:ecto-dermal interaction). In: Butler PM, Joysey KA (eds), *Development, Function and Evolution of Teeth*. London: Academic Press, pp 99–109.

Pennycuik PR, Raphael KA: 1984. The tabby locus (*Ta*) in the mouse: its site of action in tail and body skin. *Genet Res* 43: 51–63.

Roberts RC: 1966. Private communication. *Mouse News Lett* 35: 24 only.

Roderick TH, Davisson MT: 1986. Private communication. *Mouse News Lett* 75: 15 only.

Selmanowitz VJ: 1979. Ectodermal dysplasias including epitheliogenesis imperfecta, ich-thyoses, and follicular/glandular anomalies. In: Andrews EJ, Ward BC, Altman NH (eds), *Spontaneous Animal Models of Human Diseases*. New York: Academic Press, vol II, p 8.

Selmanowitz VJ, Kramer KM, Orentreich N: 1970. Congenital ectodermal defect in miniature poodles. *J Hered* 61: 196–199.

Sofaer JA: 1969a. Aspects of the tabby–crinkled–downless syndrome. I. The development of tabby teeth. *J Embryol Exp Morphol* 22: 181–205.

Sofaer JA: 1969b. Aspects of the tabby–crinkled–downless syndrome. II. Observations on the reaction to changes of genetic background. *J Embryol Exp Morphol* 22: 207–227.

Sofaer JA: 1974. Differences between *tabby* and *downless* mouse epidermis and dermis in culture. *Genet Res* 23: 219–225.

Sofaer JA: 1979. Additive effects of the genes *tabby* and *crinkled* on tooth size in the mouse. *Genet Res* 33: 169–174.

Stevens LC: 1963. Private communication. *Mouse News Lett* 29: 40 only.

Thomsett RL: 1961. Congenital hypotrichia in the dog. *Vet Rec* 73: 915–917.

Wijeratne WVS, O'Toole D, Wood L, Harkness JW: 1988. A genetic, pathological and virological study of congenital hypotrichosis and incisor anodontia in cattle. *Vet Rec* 122: 149–152.

*30536 EYE–EAR REDUCTION [?30970]

Mouse, *Ie*

This mutation, which arose following irradiation of spermatogonia, was described by Hunsicker (1974). Homozygous females and hemizygous males have anophthalmia and very small external ears with thickened, crinkled edges. The phenotype in heterozygous females may vary from normal to that of typically hemizygous or homozygous animals. The mutation does not appear to alter viability or fertility. Lyon (1974) suggested that the mutation might be homologous with an X-linked microphthalmia in man.

Hunsicker P: 1974. Private communication. *Mouse News Lett* 50: 51–52.
Lyon MF: 1974. Mechanisms and evolutionary origins of variable X-chromosome activity in mammals. *Proc R Soc Lond [Biol]* 187: 243–268.

*30549 FIDGET, X-LINKED [NK]

Mouse, *Slf*

This trait was found in a search for sex-linked recessive lethals. It is semidominant; hemizygotes show a slight up-and-down shaking of the head, which is most easily identified around weaning age; a few heterozygotes show similar behavior, but most are normal (Phillips and Fisher 1978).

Phillips RJS, Fisher G: 1978. Private communication. *Mouse News Lett* 58: 44 only.

*30554 FRAGILE SITE [NK]

Cattle

Uchida et al. (1986) described a fragile site in 3–4% of the cells of a female calf affected with the baldy calf syndrome, its normal sister and mother, and an unrelated Holstein cow. This site is probably not homologous with fra(X)(q27 or 28) in man.

Uchida IA, Freeman VCP, Basrur PK: 1986. The fragile X in cattle. *Am J Med Genet* 23: 557–562.

*30556 FRAGILE SITE fra(X)(q1), FOLATE-SENSITIVE [NK]

Indian mole rat

Tewari et al. (1987) described 4 folate-sensitive fragile sites on the X chromosome of the Indian mole rat. All normal females were heterozygous for fragile sites on the constitutive heterochromatic long arm; no fragile sites were present on the X chromosome of normal males. This finding suggests that the maternal fragile X is preferentially transmitted to daughters. The location of the sites corresponds to regions of constitutive heterochromatic deletions which result in a variety of polymorphic X chromosomes in natural populations. In individuals that did not reproduce in the laboratory, there were more fragile sites on both X chromosomes of females and on the X chromosome of males. These fragile sites differ in location, prevalence, and pattern of expression from the fra(X)(q27 or 28) associated with mental retardation in man.

Tewari R, Juyal RC, Thelma BK, Das BC, Rao SRV: 1987. Folate-sensitive fragile sites on the X-chromosome heterochromatin of the Indian mole rat, *Nesokia indica. Cytogenet Cell Genet* 44: 11–17.

*30557 FRAGILE SITE fra(X)(q2), FOLATE-SENSITIVE [NK]

See 30556.

*30558 FRAGILE SITE fra(X)(q3), FOLATE-SENSITIVE [NK]

See 30556.

*30559 FRAGILE SITE fra(X)(cen), FOLATE-SENSITIVE [NK]

See 30556.

*30561 FRAGILE SITE XC–D [NK]

Mouse

Djalali et al. (1987) induced fragile sites on murine chromosomes with the antimetabolites methotrexate, fluorodeoxyuridine, and aphidicolin. A fragile site at XC–D was by far the most sensitive. It is probably part of the *Tfm–Pgk-1–Ags* linkage group and hence homologous to the human fragile site, fra(X)(q22).

Djalali M, Adolph S, Steinbach P, Winking H, Hameister H: 1987. A comparative mapping study of fragile sites in the human and murine genomes. *Hum Genet* 77: 157–162.

*30590 GLUCOSE-6-PHOSPHATE DEHYDROGENASE (G6PD, CD 1.1.1.49)

This is the structural locus for this enzyme.

American mink

Rubtsov et al. (1981) used somatic cell hybridization to demonstrate X-linkage provisionally.

Cat

O'Brien and Nash (1982) used somatic cell hybridization to demonstrate X-linkage.

Cattle

X-linkage has been demonstrated by Heuertz and Hors-Cayla (1978), Forster et al. (1980), and Shimizu et al. (1981), using cell hybridization.

Chinese hamster

X-linkage was proved by gene dosage studies (Westerveld et al. 1972) and by the use of ovary cells deficient in G6PD and HPRT (Rosenstraus and Chasin 1975).

Dasyurids

In 2 species, *Sminthopsis crassicaudata* and *Planigale maculata*, of this marsupial superfamily, the locus is syntenic with *HPRT, GLA*, and *PGK* (Dobrovic and Graves 1986); the gene order appears to be *G6PD–HPRT–PGK–GLA*, with the first-named locus well separated from the others.

Dog

Meera Khan et al. (1973) demonstrated X-linkage by electrophoresing red blood cells; 91 dogs were of type G6PD1, but 1 male had a variant (G6PD2) with about 90% of the normal electrophoretic mobility in both red and white cells. Unfortunately, the pedigree could not be traced, and the dog had been castrated. Subsequently, Meera Khan et al. (1984) confirmed X-linkage by using somatic cell hybridization.

Donkey and horse

Donkey and horse G6PD are distinguishable on starch gel electrophoresis. Trujillo et al. (1965) and Mathai et al. (1966) used reciprocal interspecific hybrids to determine that G6PD is X-linked. References to more recent studies are given by Pearson et al. (1979).

Hare

Using reciprocal interspecific crosses of European wild hares (*Lepu europaeus* and *L. timidus*), Ohno et al. (1965) obtained evidence suggesting X-linkage.

Indian muntjac

Shows et al. (1976) demonstrated X-linkage by somatic cell hybridization, and Ropers et al. (1982) assigned the gene to the short arm.

Macropodids (kangaroos and wallabies)

Two electrophoretic forms (fast and slow) of G6PD occur in wallaroos and euros (*Macropus robustus*), and one (slow) in the red kangaroo (*M. rufus*); Richardson et al. (1971) used crosses between various species and subspecies with the different enzyme forms to demonstrate X-linkage of the gene. Johnston et al. (1975) showed that the gene is X-linked in the red-necked wallaby (*M. rufogriseus*) in which 2 alleles are present. Dosage compensation occurs through paternal X inactivation (Johnson and Sharman 1975). Graves et al. (1979) used marsupial–mouse cell hybrids to demonstrate that the genes for G6PD, HPRT, and PGK are X-linked, and Donald and Hope (1981) used hybrid cell studies to localize the 3 genes to the terminal portion of the euchromatic arm of the red kangaroo X chromosome. Dawson and Graves (1984) showed that the gene order is *G6PD–HPRT–PGK* in the gray kangaroo, and later demonstrated the same gene order on the long arm of the euro and wallaroo X chromosomes (Dawson and Graves 1986). On the long arm of the wallaroo X, *GLA* is located between *HPRT* and *G6PD*. It appears that all *Macropus* species share a common segment bearing these loci.

Mouse, *G6pd*

Epstein (1969) used the oocytes of XO and XX female mice to demonstrate that G6PD is X-linked, that its synthesis occurs in the oocyte and is dose dependent. *Mus musculus* and *M. caroli*, a related species from Thailand, have different electrophoretic mobilities for a number of enzymes including G6PD; Chapman and Shows (1976) and Shows et al. (1976) used cells from fetuses obtained by interspecific hybridization through artificial insemination to demonstrate X-linkage. Among the offspring of a male treated with ethylnitrosourea, Pretsch et al. (1988) found a mutation that causes a deficiency of G6PD. Hemizygotes, heterozygotes, and homozygotes have, respectively, about 20%, 60%, and 15% G6PD activity in the blood; this reduction in activity is also observed in extracts of other tissues. Spleen weight in hemizygotes is about twice that of normal mice; other hematologic variables are normal. The colonic epithelial cells of heterozygotes express either normal or low activity and form patches composed of multiple crypts of uniform phenotype (Griffiths et al. 1988). Peters and Ball (1985) used the low-activity mutant to locate the locus proximally between *Hq* and *Ta*. Earlier, Martin-DeLeon et al. (1985) used in situ hybridization to assign the locus to the A region. The original symbol *Gpdx* was changed to *G6pd* in 1985 (Peters 1985).

Pig

Gellin et al. (1980) demonstrated synteny with *PGK* and *HPRT*. Verhorst (1973) suggested that polymorphism exists in German Large-White swine; however, the data presented indicate that the polymorphism is autosomal. Forster et al. (1980) and Leong et al. (1983) demonstrated X-linkage by somatic cell hybridization.

Primates (nonhuman)

There seems to be much less variability of G6PD among nonhuman mammals than in man. Gourdin et al. (1972) found a rapid variant in 2 groups of baboons from Senegal; but no direct evidence of X-linkage has been published. Finaz et al. (1975) demonstrated X-linkage in the chimpanzee, which was confirmed by Chen et al. (1976) and Garver et al. (1978). Among red cells from 54 chimpanzees, Beutler and West (1978) discovered 2 electrophoretic variants, 1 of which seemed to be associated with modestly diminished enzyme activity. X-linkage has been demonstrated in the African green monkey, gorilla, orangutan, and rhesus monkey (Garver et al. 1978); in the mouse lemur (Cochet et al. 1982); and in the owl monkey (Ma 1983).

Rabbit

Echard and Gillois (1979) and Cianfriglia et al. (1979) used somatic cell hybridization to demonstrate that the locus is syntenic with *Hprt, Pgk,* and *Gla;* Echard et al. (1981) subsequently localized it on the X chromosome.

Rat

Yoshida (1978) demonstrated X-linkage. Ho et al. (1988) cloned and sequenced a cDNA encoding rat G6PD. Although the deduced amino acid sequences of 479 residues of the rat and human enzymes are 94% congruent, the amino acid terminal sequences of the 2 enzymes are not similar.

Red fox (*Vulpes vulpes*)

Rubtsov et al. (1987) demonstrated X-linkage.

Sheep

Saïdi-Mehtar et al. (1981) demonstrated synteny with *Pgk* and *Gla,* and indirectly with *Hprt.*

Virginia opossum (*Didelphis virginiana*), *Gpd*

Samollow et al. (1987) demonstrated that the locus is X-linked in this American marsupial. The paternal allele is preferentially inactivated, as it is in Australian marsupials, but dosage compensation is not complete: the gene is partially active in a large number, but not all, of tissues examined. Not all X-linked genes behave this way; see *Pgk-A* (31180).

Vole

Cook (1975) demonstrated X-linkage by cell hybridization between mouse cells and vole lymphocytes.

Beutler E, West C: 1978. Glucose-6-phosphate dehydrogenase variants in the chimpanzee. *Biochem Med* 20: 364–370.

Chapman VM, Shows TB: 1976. Somatic cell genetic evidence for X-chromosome linkage of three enzymes in the mouse. *Nature* 259: 665–667.

Chen S, McDougall JK, Creagan RP, Lewis V, Ruddle FH: 1976. Mapping of genes and adenovirus-12-induced gaps using chimpanzee–mouse somatic cell hybrids. *Cytogenet Cell Genet* 16: 412–415.

Cianfriglia M, Miggiano VC, Meo T, Muller HJ, Muller E, Battistuzzi G: 1979. Evidence for synteny between the rabbit *HPRT, PGK* and *G6PD* in mouse × rabbit somatic cell hybrids. (Abstr) *Cytogenet Cell Genet* 25: 142 only.

Cochet C, Creau-Goldberg N, Turleau C, de Grouchy J: 1982. Gene mapping of *Microcebus murinus* (Lemuridae): a comparison with man and *Cebus capucius* (Cebidae). *Cytogenet Cell Genet* 33: 213–221.

Cook PR: 1975. Linkage of the loci for glucose-6-phosphate dehydrogenase and for iosinic acid pyrophosphorylase to the X chromosome of the field-vole *Microtus agrestis. J Cell Sci* 17: 95–112.

Dawson GW, Graves JAM: 1984. Gene mapping in marsupials and monotremes. I. The chromosomes of rodent–marsupial (*Macropus*) cell hybrids, and gene assignments to the X chromosome of the grey kangaroos. *Chromosoma* 91: 20–27.

Dawson GW, Graves JAM: 1986. Gene mapping in marsupials and monotremes. III. Assignment of four genes to the X chromosome of the wallaroo and the euro (*Macropus robustus*). *Cytogenet Cell Genet* 42: 80–84.

Dobrovic A, Graves JAM: 1986. Gene mapping in marsupials and monotremes. II. Assignments to the X chromosome of dasyurid marsupials. *Cytogenet Cell Genet* 41: 9–13.

Donald JA, Hope RM: 1981. Mapping a marsupial X chromosome using kangaroo–mouse somatic cell hybrids. *Cytogenet Cell Genet* 29: 127–137.

Echard G, Gillois M: 1979. *G6PD–PGK–GLA–HPRT* synteny in the rabbit, *Oryctolagus cunniculus. Cytogenet Cell Genet* 25: 148–149.

Echard G, Gellin J, Benne F, Gillois M: 1981. The gene map of the rabbit (*Oryctolagus cuniculus* L.). I. Synteny between the rabbit gene loci coding for HPRT, PGK, G6PD, and GLA: their localization on the X chromosome. *Cytogenet Cell Genet* 29: 176–183.

Epstein CJ: 1969. Mammalian oocytes: X chromosome activity. *Science* 163: 1078–1079.

Finaz C, Cochet C, de Grouchy J, Van Cong N, Rebourcet R, Frezal J: 1975. Localisations géniques chez le chimpanze (*Pan troglodytes*). Comparaison avec la carte factorielle de l'homme (*Homo sapiens*). *Ann Génét* 18: 169–177.

Förster M, Stranzinger G, Hellkuhl B: 1980. X-chromosome gene assignment of swine and cattle. *Naturwissenschaften* 67: 48–49.

Garver JJ, Pearson PL, Estop A, Dijksman TM, Wijnen LMM, Westerveld A, Meera Khan P: 1978. Gene assignments to the presumptive homologs of human chromosomes 1, 6, 11, 12, and X in the Pongidae and Cercopithecoidea. *Cytogenet Cell Genet* 22: 564–569.

Gellin J, Benne F, Hors-Cayla MC, Gillois M: 1980. Carte génique du porc (*Sus scrofa l.*). I. Etude de deux groupes synténiques G6PD, PGK, HPRT et PKM2, MPI. *Ann Génét* 23:15–21.

Gourdin D, Vergnes H, Bouloux C, Ruffie J, Gherardi M: 1972. Polymorphism of erthrocyte G-6-PD in the baboon. *Am J Phys Anthropol* 37: 281–288.

Graves JAM, Chew GK, Cooper DW, Johnston PG: 1979. Marsupial–mouse cell hybrids containing fragments of the marsupial X chromosome. *Somatic Cell Genet* 5: 481–489.

Griffiths DFR, Davies SJ, Williams D, Williams GT, Williams ED: 1988. Demonstration of somatic mutation and colonic crypt clonality by X-linked enzyme histochemistry. *Nature* 333: 461–463.

Heuertz S, Hors-Cayla MC: 1978. Carte génétique des bovins par la technique d'hybridation cellulaire. Localisation sur le chromosome X de la glucose-6-phosphate déshydrogénase, la phosphoglycérate kinase, l'alpha-galactosidase A et l'hypoxanthine guanine phosphoribosyl transferase. *Ann Génét* 21: 197–202.

Ho Y-S, Howard AJ, Crapo JD: 1988. Cloning and sequence of a cDNA encoding rat glucose-6-phosphate dehydrogenase. *Nucleic Acids Res* 16: 7746 only.

Johnston PG, Sharman GB: 1975.Studies on metatherian sex chromosomes. I. Inheritance and inactivation of sex-linked allelic genes determining glucose-6-phosphate dehydrogenase variation in kangaroos. *Aust J Biol Sci* 28: 567–574.

Johnston PG, VandeBerg JL, Sharman GB: 1975. Inheritance of erythrocyte glucose-6-phosphate dehydrogenase in the red-necked wallaby, *Macropus rufogriseus* (Demarest), consistent with paternal X inactivation. *Biochem Genet* 13: 235–242.

Leong MML, Lin CC, Ruth RF: 1983. The localization of genes for HPRT, G6PD, and alpha-GAL onto the X-chromosome of domestic pig (*Sus scrofa domesticus*). *Can J Genet Cytol* 25: 239–245.

Ma NSF: 1983. Gene mapping of the New World Bolivian owl monkey, *Aotus. J Hered* 74: 27–33.

Martin-DeLeon PA, Wolf SF, Persico G, Toniolo D, Martini G, Migeon BR: 1985. Localization of glucose-6-phosphate dehydrogenase in mouse and man by in situ hybridization: evidence for a single locus and transposition of homologous X-linked genes. *Cytogenet Cell Genet* 39: 87–92.

Mathai CK, Ohno S, Beutler E: 1966. Sex-linkage of the glucose-6-phosphate dehydrogenase gene in Equidae. *Nature* 210: 115–116.

Meera Khan P, Los WRT, van der Does JA, Epstein RB: 1973. Isoenzyme markers in dog blood cells. *Transplantation* 15: 624–628.

Meera Khan P, Brahe C, Wijnen LMM: 1984. Gene map of dog: six conserved and three disrupted syntenies. (Abstr) *Cytogenet Cell Genet* 37: 537–538.

O'Brien SJ, Nash WG: 1982. Genetic mapping in mammals: chromosome map of domestic cat. *Science* 216: 257–265.

Ohno S, Poole J, Gustavsson I: 1965. Sex-linkage of erythrocyte glucose-6-phosphate dehydrogenase in two species of wild hares. *Science* 150: 1737–1738.

Pearson PL, Roderick TH, Davisson MT, Garver JJ, Warburton D, Lalley PA, O'Brien SJ: 1979. Report of the Committee on Comparative Mapping. *Cytogenet Cell Genet* 25: 82–95.

Peters J: 1985. Private communication. *Mouse News Lett* 73: 4 only.

Peters J, Ball SJ: 1985. Private communication. *Mouse News Lett* 73: 17–18.

Pretsch W, Charles DJ, Merkle S: 1988. X-linked glucose-6-phosphate dehydrogenase deficiency in *Mus musculus. Biochem Genet* 26: 89–103.

Richardson BJ, Czuppon AB, Sharman GB: 1971. Inheritance of glucose-6-phosphate dehydrogenase variation in kangaroos. *Nature [New Biol]* 230: 154–155.

Ropers HH, Sperling K, Raman R, Schmelzer B, Stromaier U, Neitzel H: 1982. Indian muntjac: gene assignments on the short and the long arm of the X chromosome. (Abstr) *Cytogenet Cell Genet* 32: 312 only.

Rosenstraus M, Chasin LA: 1975. Isolation of mammalian cell mutants deficient in glucose-6-phosphate dehydrogenase activity: linkage to hypoxanthine phosphoribosyl transferase. *Proc Natl Acad Sci USA* 72: 493–497.

Rubtsov NB, Radjabli SI, Gradov AA, Serov OL: 1981. Chinese hamster × American mink somatic cell hybrids: characterization of a clone panel and assignment of the mink genes for malate dehydrogenase, NADP-1 and malate dehydrogenase, NAD-1. *Theor Appl Genet* 60: 90–106.

Rubtsov NB, Matveeva VG, Radjabli SI, Kulbakina NA, Nesterova TB, Zakian SM: 1987. Construction of a clone panel of fox–Chinese hamster somatic cell hybrids and assignment of genes for LDHA, LDHB, GPI, ESD, G6PD, HPRT, alpha-GALA in the silver fox. (Russ, Eng summary) *Genetika* 23: 1088–1096.

Saïdi-Mehtar N, Hors-Cayla MC, Van Cong N: 1981. Sheep gene mapping by somatic cell hybridization: four syntenic groups: ENO1–PGD, ME1–PGM3, LDHB–PEPB–TPI, and G6PD–PGK–GALA. *Cytogenet Cell Genet* 30: 193–204.

Samollow PB, Ford AL, VandeBerg JL: 1987. X-linked gene expression in the Virginia opossum: differences between the paternally derived *Gpd* and *Pgk-A* loci. *Genetics* 115: 185–195.

Shimizu N, Shimizu Y, Kondo I, Woods C, Wenger T: 1981. The bovine genes for phosphoglycerate kinase, glucose-6-phosphate dehydrogenase, alpha-galactosidase, and hypoxanthine phosphoribosyltransferase are linked to the X chromosome in cattle–mouse hybrids. *Cytogenet Cell Genet* 29: 26–31.

Shows TB, Brown JA, Chapman VM: 1976. Comparative gene mapping of *HPRT, G6PD,* and *PGK* in man, mouse and muntjac deer. *Cytogenet Cell Genet* 16: 436–439.

Trujillo JM, Walden B, O'Neil P, Anstall HB: 1965. Sex-linkage of glucose-6-phosphate dehydrogenase in the horse and donkey. *Science* 148: 1603–1604.

Verhorst D: 1973. Polymorphism in glucose-6-phosphate dehydrogenase in the German Large-White. *Anim Blood Groups Biochem Genet* 4: 65–68.

Westerveld A, Visser RPLS, Freeke MA, Bootsma D: 1972. Evidence for linkage of 3-phosphoglycerate kinase, hypoxanthine–guanine phosphoribosyl transferase, and glucose 6-phosphate dehydrogenase loci in Chinese hamster cells studied by using a relationship between gene multiplicity and enzyme activity. *Biochem Genet* 7: 33–40.

Yoshida MC: 1978. Rat gene mapping by rat–mouse somatic cell hybridization and a comparative Q-banding analysis between rat and mouse chromosomes. *Cytogenet Cell Genet* 22: 606–609.

30615 GONAD-SPECIFIC RECEPTOR OF H–Y ANTIGEN [NK]

Wood Lemming (*Myopus schisticolor*)

Two different X chromosomes, X and X*, occur in the wood lemming. The X* is morphologically and functionally different from the X because of rearrangements, presumably including a deletion, in the short arm. This chromsome is generally associated with female development; X*Y individuals are female. See H–Y REGULATOR (30697) for details. The finding that X*Y and X*O individuals are H–Y+ and possess ovaries (Wiberg et al. 1982) suggests that the deleted portion of the X* chromosome possessed a gene determining a gonad-specific receptor of the H–Y antigen.

Wiberg U, Mayerova A, Muller U, Fredga K, Wolf U: 1982. X-linked genes of the H–Y antigen system in the wood lemming (*Myopus schisticolor*). *Hum Genet* 60: 163–166.

*30645 GREASY [NK]

Mouse, *Gs*

This X-linked dominant trait, first described by Larsen (1964) and Russell and Larsen (1965), involves a generalized alteration in hair structure. The hairs of hemizygotes differ from those of wild-type animals in several ways: Many more lack a subterminal yellow band, almost all have very long "unsegmented tips," their average number of constrictions is reduced, they have a marked irregularity of outline, and they have a reduced number of rows of medullary cells perpendicular to the long axis. Homozygous females resemble hemizygous males in these respects, and heterozygotes are intermediate between *Gs/Gs* and *Gs/Y* on the one hand and +/+ on the other (Dunn 1972). The locus is located in midchromosome about 5 units proximally to *Mo* (Grahn 1972).

Dunn GR: 1972. Expression of a sex-linked gene in standard and fusion-chimeric mice. *J Exp Zool* 181: 1–16.
Grahn D: 1972. Private communication. *Mouse News Lett* 47: 20 only.
Larsen MM: 1964. Private communication. *Mouse News Lett* 30: 47 only.
Russell LB, Larsen MM: 1965. Private communication. *Mouse News Lett* 33: 69 only.

*30653 HARLEQUIN [NK]

Mouse

This mutation arose spontaneously in outbred CF1 stock. Hemizygous males and homozygous females are almost completely bald; both are fully viable and fertile, although they are much smaller than normal. The heterozygous females have bald patches with varying distribution (Barber 1971). These patches cover very much less than 50% of the total surface area and are not always easy to see (Falconer and

Isaacson 1972). The locus is closely linked to *Hprt* (30800) at the distal end of the proximal third of the chromosome.

Barber BR: 1971. Private communication. *Mouse News Lett* 45: 34–35.
Falconer DS, Isaacson JH: 1972. Private communication. *Mouse News Lett* 47: 28 only.

*30654 HARVEY SARCOMA PROTOONCOGENE-2 PSEUDOGENE (HRAS2; HARVEY SARCOMA CELLULAR ONCOGENE-2) [31099]

Cat

O'Brien (1986) mentions unpublished work by himself and Chang which localizes this pseudogene to the X chromosome.

Rat, c-*Hras-2*

Szpirer et al. (1985) used somatic cell hybridization to demonstrate X-linkage. The human homolog is also X-linked (O'Brien et al. 1983). The rat gene lacks intervening sequences, as does its human counterpart; hence, both are probably pseudogenes.

O'Brien SJ: 1986. Molecular genetics in the domestic cat and its relatives. *Trends Genet* 2: 137 142.
O'Brien SJ, Nash WG, Goodwin JL, Lowy DR, Change EH: 1983. Dispersion of the *ras* family of transforming genes to four different chromosomes in man. *Nature* 302: 839–842.
Szpirer J, Defeo-Jones D, Ellis RW, Levan G, Szpirer C: 1985. Assignment of three rat cellular *ras* oncogenes to chromosomes 1, 4, and X. *Somatic Cell Mol Genet* 11: 93–97.

*30655 HEMOGLOBIN PRODUCTION, INHIBITION OF [?30130]

Mouse

Benoff and Skoultchi (1977) used somatic cell hybrids between erythroleukemic cells, which have a low basal level of erythroid differentiation, and mouse lymphoma or bone marrow cells to demonstrate the existence of an X-linked locus (loci) capable of inhibiting DSMO-inducible hemoglobin production of the erythroleukemic cell. McKusick (1986) suggests possible homology with sideroblastic anemia (MIM 30130) in man.

Benoff S, Skoultchi AI: 1977. X-linked control of hemoglobin production in somatic hybrids of mouse erythroleukemic cells and mouse lymphoma or bone marrow cells. *Cell* 12: 263–274.
McKusick VA: 1986. *Mendelian Inheritance in Man: Catalogs of Autosomal Dominant, Autosomal Recessive, and X-Linked Phenotypes*, 7th ed. Baltimore: Johns Hopkins Univ Press.

*30692 HINDLEG PARALYSIS [?30280]

Syrian hamster, *pa*

Hindleg paralysis occurs as a recessive trait in the Syrian hamster (Nixon and Conneally 1968). The defect appears between 6 and 10 months of age and exhibits considerable variability in rate of progress of the paralysis, and severity of the final stage. Although this variation occurs in both hemizygous males and homozygous females, the expression in the latter is generally not as severe, and there is nonpenetrance in some females. In its severest form, the legs are extended rigidly back from the body and the soles of the feet are often turned inward and upward. Although affected animals cannot breed after becoming paralytic, they manage to keep mobile and most live to an advanced age. Electron microscope studies of the spinal roots and sciatic nerves of affected animals (Hirano and Dembitzer 1976) reveal fibrillary accumulations, some with a crystalloid organization, in the inner loops of the myelin sheath. These changes are present to a slight degree in control animals but are severe in *pa* hemizygotes. Further studies (Hirano 1978) related these findings in the latter to demyelination, with apparent sparing of the Schwann cell perikaryon, and subsequent remyelination. In animals at advanced stages of the disease, large-caliber axons are completely denuded of myelin and are surrounded by Schwann cells alone. The primary focus of the pathogenesis is not clear. Hirano (1977) described severe axonal changes, confined to large axons, that were usually associated with an intact myelin sheath. However, the same investigator (Hirano 1978) has observed the reverse: an apparently normal intact axon surrounded by a severely damaged sheath. Additional studies by Hirano (1980) and Hirano and Dembitzer (1981) have not led to a clear delineation of the primary defect. Possible human homologies are spastic paraplegia (MIM 31290), spinal and bulbar muscular atrophy (MIM 31320) (Lyon 1974), and Charcot–Marie–Tooth peroneal muscular atrophy (MIM 30280) (Lyon 1974). On the basis of present evidence, the last seems the most reasonable.

Hirano A: 1977. Fine structural changes in the mutant hamster with hind leg paralysis. *Acta Neuropathol (Berl)* 39: 225–230.

Hirano A: 1978. A possible mechanism of demyelination in the Syrian hamster with hindleg paralysis. *Lab Invest* 38: 115–121.

Hirano A: 1980. Further observations on peripheral neuropathy in the Syrian hamster with hind limb paralysis. *Acta Neuropathol (Berl)* 50: 187–192.

Hirano A, Dembitzer HM: 1976. Eosinophilic rod-like structures in myelinated fibers of hamster spinal roots. *Neuropathol Appl Neurobiol* 2: 225–232.

Hirano A, Dembitzer HM: 1981. The periaxonal space in an experimental model of neuropathy: the mutant Syrian hamster with hindleg paralysis. *J Neurocytol* 10: 261–269.

Lyon MF: 1974. Mechanisms and evolutionary origins of variable X-chromosome activity in mammals. *Proc R Soc Lond [Biol]* 187: 243–268.

McKusick VA: 1986. *Mendelian Inheritance in Man: Catalogs of Autosomal Dominant, Autosomal Recessive, and X-Linked Phenotypes*, 7th ed. Baltimore: Johns Hopkins Univ Press.

Nixon CW, Conneally ME: 1968. Hind-leg paralysis: a new sex-linked mutation in the Syrian hamster. *J Hered* 59: 276–278.

*30694 HISTOCOMPATIBILITY–X [NK]

Mouse, *H–X*

Although Strong (1929) suggested that an X-linked histocompatibility gene existed in the mouse, Bailey (1963a,b) was the first to demonstrate clearly its existence. This finding was confirmed by Rosenau and Horwitz (1968). These studies showed that the BALB/c and C57BL/6 strains possess different alleles. Hildemann and Cooper (1967) and Hildemann et al. (1974) demonstrated that strains A and C57BL/6 possess different alleles; Berryman and Silver (1979) showed that strains A and BALB/c bear different alleles and that DBA/2 animals carry yet another allele. The alleles of B6 and BALB have been designated $H–X^b$ and $H–X^c$, respectively (Graff 1979); and those of A and DBA/2 have been suggested as $H–X^a$ and $H–X^d$ (Berryman and Silver 1979). In a brief note, Bailey and Chai (1973) state that the LG/CKe strain carries another allele, designated $H–X^e$ (Graff 1979).

Rat, *H–X*

Mullen and Hildeman (1972) demonstrated the existence of this locus in the rat.

Bailey DW: 1963a. Histocompatibility associated with the X chromosome in mice. *Transplantation* 1: 70–74.

Bailey DW: 1963b. Mosaic histocompatibility of skin grafts from female mice. *Science* 141: 631–633.

Bailey DW, Chai LC: 1973. Allelic forms of X- linked histocompatibility. *44th Annual Report, The Jackson Laboratory, 1972–1973.* p 30 only.

Berryman PL, Silvers WK: 1979. Studies on the *H–X* locus of mice. I. Analysis of polymorphism. *Immunogenetics* 9: 363–367.

Graff RJ: 1979. Histocompatibility systems, except *H-2*: mouse. In: Altman PL, Katz DD (eds), *Inbred and Genetically Defined Strains of Laboratory Animals, Part 1: Mouse and Rat.* Bethesda MD: Federation of American Societies for Experimental Biology, pp 118–122.

Hildemann WH, Cooper EL: 1967. Transplantation genetics: unexpected histoincompatibility associated with skin grafts from F_2 and F_3 hybrid donors to F_1 hybrid recipients. *Transplantation* 5: 707–720.

Hildemann WH, Mullen Y, Inai M: 1974. Anergy to dual H–X and X–Y antigens occurring in the same skin allografts between reciprocal F_1 hybrid male mice. *Immunogenetics* 1: 297–303.

Mullen Y, Hildemann WH: 1972. X- and Y-linked transplantation antigens in rats. *Transplantation* 13: 521–529.

Rosenau W, Horwitz C: 1968. Graft rejections in paternal to F_1 hybrid and reciprocal hybrid grafts indicating a histocompatibility gene on the mouse X chromosome. *Lab Invest* 18: 298–303.

Strong LC: 1929. Transplantation studies on tumors arising spontaneously in heterozygous individuals. *J Cancer Res* 13: 103–115.

*30697 H–Y REGULATOR (HYR)

Wood lemming (*Myopus schisticolor*)

In the wood lemming, the sex ratio is 0.20–0.30, and 2 types of females occur, 1 of which produces female offspring only (Kalela and Oksala 1966). Fredga et al. (1976) reported the existence of 2 types of fertile females with different sex chromosome constitutions, XX and XY; although the XY females are anatomically normal and indistinguishable from the XX females, they produce X-type eggs only, presumably as a consequence of selective nondisjunction in the fetal ovary. Fredga et al. (1976) suggested that the XY females possess an X-linked gene that suppresses the male-determining effect of the X chromosome; the X chromosome with the mutant gene is denoted X*. This hypothesis was supported by Wachtel et al. (1976), who observed that all female wood lemmings were H–Y⁻; Fredga et al. (1977) published additional breeding and chromosome data in support of their hypothesis. Herbst et al. (1978) demonstrated that the X and X* are morphologically different: the latter is slightly shorter, presumably as a consequence of a deletion associated with structural rearrangements in the short arm. The X* generally results in the development of femaleness even in the presence of the Y chromosome. Wiberg et al. (1982) tested XY males and XX, X*Y, X*O, X*X, and XO females for the presence and amount of X–Y antigen. Only XX individuals were negative; XY, X*Y, and X*O individuals were positive, and X*X, and XO females typed in the intermediate range of the H–Y antigen titer. These serological observations were confirmed by Wiberg and Günther (1985), who found that female lemmings with the X* chromosome carry the H–Y transplantation antigen. The finding that the X*Y female is X–Y⁺ disagrees with that of Wachtel et al. (1976) mentioned above. Wiberg et al. (1982) explain these results as follows: The X–Y regulatory gene is deleted from the X*; the intermediate levels of H–Y antigen in X*X and XO females result from a single dose of the X suppressor (since the X-linked regulatory gene exhibits a gene-dosage effect, it must escape X-inactivation); and the development of a female phenotype in the presence of normal levels of X–Y antigen results from the absence of an X-linked, gonad-specific receptor. See GONAD-SPECIFIC RECEPTOR OF H–Y ANTIGEN (30615).

Fredga K, Gropp A, Winking H, Frank F: 1976. Fertile XX- and XY-type females in the wood lemming *Myopus schisticolor. Nature* 261: 225–227.

Fredga K, Gropp A, Winking H, Frank F: 1977. A hypothesis explaining the exceptional sex ratio in the wood lemming. (*Myopus schisticolor*). *Hereditas* 85: 101–104.

Herbst EW, Fredga K, Frank F, Winking H, Gropp A: 1978. Cytological identification of the two X-chromosome types in the wood lemming (*Myopus schisticolor*). *Chromosoma* 69: 185–191.

Kalela O, Oksala T: 1966. Sex ratio in the wood lemming, *Myopus schisticolor* (LILLJEB.), in nature and in captivity. *Ann Univ Turkuensis [Ser. AII]* 37: 1–24.

Wachtel SS, Koo GC, Ohno S, Gropp A, Dev VG, Tantrahavi R, Miller DA, Miller OJ: 1976. H–Y antigen and the origin of the XY female wood lemmings (*Myopus schisticolor*). *Nature* 264: 638–639.

Wiberg UH, Günther E: 1985. Female wood lemmings with the mutant X*-chromosome carry the H–Y transplantation antigen. *Immunogenetics* 21: 91–96.

Wiberg U, Mayerová A, Müller U, Fredga K, Wolf U: 1982. X-linked genes of the H–Y antigen system in the wood lemming (*Myopus schisticolor*). *Hum Genet* 60: 163–166.

30702 HYDROCEPHALUS, X-LINKED [?30700]

Rat, *Hyd*

The first animals with this disorder were observed in the Csk:Wistar–Imamichi strain in Japan (Koto et al. 1987). The hydrocephalus is severe in 96% of males; affected animals can be distinguished about 7 days after birth by their dome-shaped head, shortened nose, and slow growth. These males become inactive, and some develop ataxia and convulsions; almost all of them die within the first month. The key features at autopsy are marked dilation of the ventricles (particularly the lateral ones), severe thinning of the cerebral mantle, and a malformed midbrain. The ventricles are developmentally normal and the aqueduct is patent. Some females are severely affected, but most (87%) are only mildly affected; they are externally normal, grow well, and live to maturity. The features noted at autopsy in the males are present in a much milder form. Although the authors state that breeding data indicate the disorder to be a dominant X-linked trait, the only data presented are difficult to interpret, and no conclusions can be made now about the genetics. An X-linked form of hydrocephalus is known in man (MIM 30700); however, the clinical and neuropathologic picture is not clear, and, given the confused state of the genetics in the rat disorder, a firm statement about homology would be premature at present.

Koto M, Miwa M, Shimizu A, Tsuji K, Okamoto M, Adachi J: 1987. Inherited hydrocephalus in Csk:Wistar–Imamichi rats; Hyd strain: a new disease model for hydrocephalus. *Exp Anim* 36: 157–162.

*30780 HYPOPHOSPHATEMIA, X-LINKED, TYPE I (VITAMIN D-RESISTANT RICKETS, X-LINKED; HYPOPHOSPHATEMIC D-RESISTANT RICKETS I)

Mouse, *Hyp*

Eicher et al. (1976) described a mutant house mouse that appears to be a precise homology of the human disorder on the basis of the nature and relative time of appearance of its manifestations; the mutation appeared to be spontaneous and is maintained on C57BL/6J. Affected animals have hypophosphatemia, bone changes resembling rickets, diminished bone ash, dwarfism, and fractional excretion of phosphate anion (Meyer RA et al. 1979). Craniofacial and dental anomalies (Iorio et al. 1979a,b 1980; Mostafa et al. 1982; Sofaer and Southam 1982) and craniosynostosis (Roy et al. 1981) are also present. The skeletal abnormalities are generally less severe in heterozygous females; the reason for this difference is not clear (Meyer RA 1985). Phosphate supplementation of the diet from weaning age prevents severe skeletal defects. In Hyp/Y mice, the renal handling of phosphate is impaired (Giasson et al. 1977; Cowgill et al. 1979). The defect in transcellular transport of phosphate in the nephron of hemizygotes is located in the brush-border membranes (Tenenhouse and Scriver 1978, 1979; Tenenhouse et al. 1978). Mühlbauer et al. (1982) suggested that there is a defect in the mechanism responsible for adapting the tubular P_i transport system to P_i restriction. The finding that there is a primary disorder of intestinal P_i transport in Hyp/Y mice (O'Doherty et al. 1977) was not confirmed by Tenenhouse et al. (1981); Tenenhouse and Scriver (1981) found that the 1,25-dihydroxyvitamin D_3 [1,25-$(OH)_2D_3$] enhances intestinal absorption of P_i and elevates serum P_i to normal levels without correcting the renal P_i transport defect. Studies in vitro (Brunette et al. 1979; Cowgill et al. 1979) and in vivo (Kiebzak and Meyer 1982b) demonstrated that the kidneys of Hyp/Y mice are not hypersensitive to parathyroid hormone. Hyp/Y mice are not hypersensitive to calcitonin (Kiebzak and Meyer 1982a). Both 1,25-$(OH)_2D_3$ and 1alpha-hydroxyvitamin D_3 are able to increase urinary phosphate conservation and improve rachitic bone morphology, but only the latter repairs the hypophosphatemia and significantly increases intestinal transport of phosphate (Beamer et al. 1980). In affected mice, plasma 1,25-$(OH)_2D_3$ levels are not increased by a low-phosphate challenge (Meyer RA et al. 1980) but are increased by a low-calcium challenge (Meyer RA et al. 1982). Phosphate supplementation heals the defective epiphyseal calcification but not the defective bone mineralization (Marie et al. 1981); no better results were obtained when 1,25- $(OH)_2D_3$ or 24,25-dihydroxy-vitamin D_3 was infused continuously for 3 weeks in 137-day-old affected mice receiving phosphate supplementation (Marie et al. 1982a). However, 4 weeks of treatment with 1,25-$(OH)_2D_3$ in postweaning animals resulted in the healing of both rachitic and osteomalacic bone lesions (Marie et al. 1982b). Tenenhouse (1983) demonstrated the presence of an abnormal 25-hydroxyvitamin D_3-1-hydroxylase (1-OHase) response in renal mitochondria; the underlying mechanism for this abnormal response and its relation to the renal brush-border membrane phosphate defect is undetermined. Renal homogenates from hypophosphatemic mice synthesize significantly less 1,25-$(OH)_2D_3$ than homogenates from normal mice with comparable hypophosphatemia produced by phosphate deprivation (Lobaugh and Drezner 1983). The inappropriate activation of 1-OHase in these animals may result from an abnormal regulatory step subsequent to the increased renal production of cAMP (Tenenhouse 1984). cAMP-dependent protein kinase and protein kinase inhibitor activity are normal in Hyp/Y kidney cytosol, but protein kinase C activity is significantly elevated in affected mice; the activity is not increased in other tissues (Tenenhouse and Henry 1985). These results suggest that high protein-kinase C activity is specific to kidney from affected mice and may be related to the renal defects in brush-border membrane phosphate transport and mitochondrial vitamin D metabolism. The kidney defect is intrinsic to renal cells: Bell et al. (1986) demonstrated the expression of the mutation in cultured renal epithelial cells from Hyp/Y mice. $Hyp/+$ females accumulate normal amounts of phosphate in their milk (Delzer and Meyer 1983). Intestinal levels of alkaline phosphatase and total intestinal protein are comparable at all ages to those in normal mice; however, the concentration of intestinal calcium-binding protein (CaBP) is significantly decreased during the weaning period, is further depressed during the period of rapid growth, and is maintained at a level about 70% of normal in adults (Bruns et al. 1984). Young affected mice, in which the levels of duodenal vitamin D-dependent CaBP are low, have low intestinal absorption of Ca and low skeletal mineralization; this dysfunction exacerbates the bone disease (Meyer MH et al. 1984a; Meyer RA et al. 1986), and affected juvenile males fail to accumulate Ca in the femur and whole body (Kay et al. 1985). Meyer RA et al. (1987) showed that low plasma $1,25\text{-}(OH)_2D_3$ causes the intestinal malabsorption of calcium and phosphate in affected juvenile mice. Yamamoto et al. (1988a) found that the serum phosphate concentration and nuclear $1,25\text{-}(OH)_2D_3$ uptake by duodenal mucosal cells increased significantly in affected mice given a high-phosphate diet. The increased food intake by $Hyp/+$ mice (Meyer MH et al. 1984b) can be explained by their increased metabolic rate (Vaughn et al. 1986). It appears that affected animals have normal levels of the metabolites of vitamin D (Meyer RA et al. 1984); however, this conclusion contradicts some findings (Cunningham et al. 1983; Brazy and Drezner 1983). Bar-Shavit et al. (1983) demonstrated defective binding of macrophages to bone in vitro; this defect appears to be attributable to the absence or inaccessibility of bone matrix oligosaccharides or glycoproteins essential to the attachment process. In comparing the mouse and human forms of the disease, Meyer RA (1985) points out that the bone disease is worse in mice. Kiebzak and Dousa (1985) demonstrated that parenterally administered thyroid hormones, T_3 or T_4, increased the renal brush-border membrane transport of P_i, decreased renal P_i excretion, and increased plasma P_i in affected animals. Osteoblast cultures from affected mice failed to respond – as measured by alkaline phosphates activity and cell proliferation – to a physiologic dose of $1,25\text{-}(OH)_2D_3$ (Yamamoto et al. 1988b). The kidney brush-border membrane vesicles of the mouse kidney have 2 distinct Na^+-dependent phosphate transport systems that differ in their affinities and capacities. Although both systems can be modulated by phosphonoformic acid and dietary phosphate, only the high-affinity one is altered in affected mice (Tenenhouse et al. 1988). *Gyro*, a mutation at a closely linked locus, also results in hypophosphatemia (see 30781). The locus is closely linked to *Li* (30898) (Cattanach 1985) at the distal end of the chromosome.

Bar-Shavit Z, Kahn AJ, Teitelbaum SL: 1983. Defective binding of macrophages to bone in rodent osteomalacia and vitamin D deficiency: in vitro evidence for a cellular defect and altered saccharides in the bone matrix. *J Clin Invest* 72: 526–534.

Beamer WG, Wilson MC, DeLuca HF: 1980. Successful treatment of genetically hypophosphatemic mice by 1alpha-hydroxyvitamin D_3 but not 1,25-dihydroxyvitamin D_3. *Endocrinology* 106: 1949–1955.

Bell CL, Scriver CR, Tenenhouse HS: 1986. Expression of the phosphate transport mutation in cultured renal epithelial cells from X-linked hypophosphatemic (*Hyp*) mice. (Abstr) *Am J Hum Genet* 39: A5 only.

Brazy PC, Drezner MK: 1983. Modulation of phosphate transport in mouse renal tubules by vitamin D metabolites: a potential mechanism for renal phosphate wasting in X-linked hypophosphatemic (Hyp) mice. (Abstr) *Calcif Tissue Int* 35: 666 only.

Brunette MG, Chabardes D, Imbert-Teboul M, Clique A, Montegut M, Morel F: 1979. Hormone-sensitive adenylate cyclase along the nephron of genetically hypophosphatemic mice. *Kidney Int* 15: 357–369.

Bruns ME, Meyer RA Jr, Meyer MH: 1984. Low levels of intestinal vitamin D-dependent calcium-binding protein in juvenile X-linked hypophosphatemic mice. *Endocrinology* 115: 1459–1463.

Cattanach BM: 1985. Private communication. *Mouse News Lett* 73: 17 only.

Cowgill LD, Goldfarb S, Lau K, Slatopolsky E, Agus ZS: 1979. Evidence for an intrinsic renal tubular defect in mice with genetic hypophosphatemic rickets. *J Clin Invest* 63: 1203–1210.

Cunningham J, Gomes H, Seino Y, Chase LR: 1983. Abnormal 24-hydroxylation of 25-hydroxyvitamin D in the X-linked hypophosphatemic mouse. *Endocrinology* 112: 633–638.

Delzer PR, Meyer RA Jr: 1983. Normal milk composition in lactating X-linked hypophosphatemic mice despite continued hypophosphatemia. *Calcif Tissue Int* 35: 750–754.

Eicher EM, Southard JL, Scriver CR, Glorieux FH: 1976. Hypophosphatemia: mouse model for human familial hypophosphatemic (vitamin D-resistant) rickets. *Proc Natl Acad Sci USA* 73: 4667–4671.

Giasson SD, Brunette MG, Danan G, Vigneault N, Carriere S: 1977. Micropuncture study of renal phosphorus transport in hypophosphatemic vitamin D resistant rickets mice. *Pflugers Arch* 371: 33–38.

Iorio RJ, Bell WA, Meyer MH, Meyer RA Jr: 1979a. Radiographic evidence of craniofacial and dental abnormalities in the X-linked hypophosphatemic mouse. *Ann Dent* 38: 31–37.

Iorio RJ, Bell WA, Meyer MH, Meyer RA Jr: 1979b. Histologic evidence of calcification abnormalities in teeth and alveolar bone of mice with X-linked dominant hypophosphatemia (VDRR). *Ann Dent* 33: 38–44.

Iorio RJ, Murray G, Meyer RA Jr: 1980. Craniometric measurements of craniofacial malformation in mice with X-linked, dominant hypophosphatemia (vitamin D-resistant rickets). *Teratology* 22: 291–298.

Kay MA, Meyer MH, Delzer PR, Meyer RA Jr: 1985. Changing patterns of femoral and skeletal mineralization during growth in juvenile X-linked hypophosphatemic mice. *Miner Electrolyte Metab* 11: 374–380.

Kiebzak GM, Douse TP: 1985. Thyroid hormones increase renal brush border membrane transport of phosphate in X-linked hypophosphatemic (*Hyp*) mice. *Endocrinology* 117: 613–619.

Kiebzak GM, Meyer RA Jr: 1982a. In vivo response of X-linked hypophosphatemic mice to calcitonin. *Horm Metab Res* 14: 216–220.

Kiebzak GM, Meyer RA Jr: 1982b. X-linked hypophosphatemic mice are not hypersensitive to parathyroid hormone. *Endocrinology* 110: 1030–1036.

Lobaugh B, Drezner MK: 1983. Abnormal regulation of renal 25-hydroxy-vitamin D-1alpha-hydroxylase activity in the X-linked hypophosphatemic mouse. *J Clin Invest* 71: 400–403.

Marie PJ, Travers R, Glorieux FH: 1981. Healing of rickets with phosphate supplementation in the hypophosphatemic male mouse. *J Clin Invest* 67: 911–914.

Marie PJ, Travers R, Glorieux FH: 1982a. Bone response to phosphate and vitamin D metabolites in the hypophosphatemic male mouse. *Calcif Tissue Int* 34: 158–164.

Marie PJ, Travers R, Glorieux FH: 1982b. Healing of bone lesions with 1,25-dihydroxyvitamin D$_3$ in the young X-linked hypophosphatemic male mouse. *Endocrinology* 111: 904–911.

Meyer MH, Meyer RA Jr, Iorio RJ: 1984a. A role for the intestine in the bone disease of juvenile X-linked hypophosphatemic mice: malabsorption of calcium and reduced skeletal mineralization. *Endocrinology* 115: 1464–1470.

Meyer MH, Meyer RA Jr, Pollard BD, Theys RD: 1984b. Abnormal trace mineral metabolism in adult X-linked hypophosphatemic mice: a possible role of increased food intake. *Miner Electrolyte Metab* 10: 1–4.

Meyer RA Jr: 1985. X-linked hypophosphatemia (familial or sex-linked vitamin-D-resistant rickets): X-linked hypophosphatemic (Hyp) mice. *Am J Pathol* 118: 340–342.

Meyer RA Jr, Jowsey J, Meyer MH: 1979. Osteomalacia and altered magnesium metabolism in the X-linked hypophosphatemic mouse. *Calcif Tissue Int* 27: 19–26.

Meyer RA Jr, Gray RW, Meyer MH: 1980. Abnormal vitamin D metabolism in the X-linked hypophosphatemic mouse. *Endocrinology* 107: 1577–1581.

Meyer RA Jr, Gray RW, Roos BA, Kiebzak GM: 1982. Increased plasma 1,25-dihydroxy-vitamin D after low calcium challenge in X-linked hypophosphatemic mice. *Endocrinology* 111: 174–177.

Meyer RA Jr, Meyer MH, Gray RW: 1984. Metabolites of vitamin D in normal and X-linked hypophosphatemic mice. *Calcif Tissue Int* 36: 662–667.

Meyer RA Jr, Meyer MH, Erickson PR, Korkor AB: 1986. Reduced absorption of [45]calcium from isolated duodenal segments *in vivo* in juvenile but not adult X-linked hypophosphatemic mice. *Calcif Tissue Int* 38: 95–102.

Meyer RA Jr, Meyer MH, Gray RW, Bruns ME: 1987. Evidence that low plasma 1,25-dihydroxyvitamin D causes intestinal malabsorption of calcium and phosphate in juvenile X-linked hypophosphatemic mice. *J Bone Min Res* 2: 67–82.

Mostafa YA, El-Mangoury NH, Meyer RA Jr, Iorio RJ: 1982. Deficient nasal bone growth in the X-linked hypophosphataemic (*Hyp*) mouse and its implication in craniofacial growth. *Arch Oral Biol* 27: 311–317.

Mühlbauer RC, Bonjour J-P, Fleisch H: 1982. Abnormal tubular adaptation to dietary P$_i$ restriction in X-linked hypophosphatemic mice. *Am J Physiol* 242: F353–F359.

O'Doherty PJA, DeLuca HF, Eicher EM: 1977. Lack of effect of vitamin D and its metabolites on intestinal phosphate transport in familial hypophosphatemia of mice. *Endocrinology* 101: 1325–1330.

Roy WA, Iorio RJ, Meyer GA: 1981. Craniosynostosis in vitamin D-resistant rickets: a mouse model. *J Neurosurg* 55: 265–271.

Sofaer JA, Southam JC: 1982. Naturally-occurring exposure of the dental pulp in mice with inherited hypophosphataemia. *Arch Oral Biol* 27: 701–703.

Tenenhouse HS: 1983. Abnormal renal mitochondrial 25-hydroxyvitamin D$_3$-1-hydroxylase activity in the vitamin D and calcium deficient X-linked *Hyp* mouse. *Endocrinology* 113: 816–818.

Tenenhouse HS: 1984. Investigation of the mechanism for abnormal renal 25-hydroxy vitamin D$_3$-1-hydroxylase activity in the X-linked *Hyp* mouse. *Endocrinology* 115: 634–639.

Tenenhouse HS, Henry HL: 1985. Protein kinase activity and protein kinase inhibitor in mouse kidney: effect of the X-linked *Hyp* mutation and vitamin D status. *Endocrinology* 117: 1719–1726.

Tenenhouse HS, Scriver CR: 1978. The defect in transcellular transport of phosphate in the nephron is located in brush-border membranes in X-linked hypophosphatemia (*Hyp* mouse model). *Can J Biochem* 56: 640–646.

Tenenhouse HS, Scriver CR: 1979. Renal brush border membrane adaptation to phosphorus deprivation in the *Hyp*/Y mouse. *Nature* 281: 225–227.

Tenenhouse HS, Scriver CR: 1981. Effect of 1,25-dihydroxyvitamin D_3 on phosphate homeostasis in the X-linked hypophosphatemic (*Hyp*) mouse. *Endocrinology* 109: 658–660.

Tenenhouse HS, Scriver CR, McInness RR, Glorieux FH: 1978. Renal handling of phosphate *in vivo* and *in vitro* by the X-linked hypophosphatemic male mouse: evidence for a defect in the brush border membrane. *Kidney Int* 14: 236–244.

Tenenhouse HS, Fast DK, Scriver CR, Koltay M: 1981. Intestinal transport of phosphate anion is not impaired in the *Hyp* (hypophosphatemic) mouse. *Biochem Biophys Res Commun* 100: 537–543.

Tenenhouse HS, Klugerman AH, Neal JL: 1988. Effect of phosphonoformic acid, dietary phosphate and the *Hyp* mutation on kinetically distinct phosphate transport processes in mouse kidney. (Abstr) *J Bone Miner Res* 3 (Suppl): S132 only.

Vaughn LK, Meyer RA Jr, Meyer MH: 1986. Increased metabolic rate in X-linked hypophosphatemic mice. *Endocrinology* 118: 441–445.

Yamamoto T, Seino Y, Tanaka H, Yamaoka K, Kurose H, Ishida M, Yabuuchi H: 1988a. Effects of the administration of phosphate on nuclear 1,25-dihydroxyvitamin D_3 uptake by duodenal mucosal cells of Hyp mice. *Endocrinology* 122: 576–580.

Yamamoto T, Ecarot-Charrier B, Glorieux FH: 1988b. Abnormal response of osteoblasts from Hyp mice to a physiological dose of 1,25-$(OH)_2D_3$. (Abstr) *J Bone Miner Res* 3 (Suppl): S221 only.

*30781 HYPOPHOSPHATEMIA, X-LINKED, TYPE II (HYPOPHOSPHATEMIC D-RESISTANT RICKETS II)

Mouse, gyro, *Gy*

The original mutant with this trait was a circling female whose dam had been exposed to x-rays in utero (Carter et al. 1960; Lyon 1960). This abnormal female produced sons that were severely affected and 3 daughters (of 28) that were mildly affected. This result suggested X-linkage, which was proved by using known markers (Lyon 1966). Hemizygous males circle, have abnormalities of the long bones and ribs, and are sterile; heterozygous females show incomplete penetrance of the circling, but are otherwise normal (Lyon 1961). Penetrance in *Gy*/+ females is partly controlled by the *Xce* locus, which is located about 5 map units distally (Cattanach et al. 1969). Hemizygous males and heterozygous females have hypophosphatemia and a rachitic type of defective bone growth similar to that in mice carrying *Hyp* (30780) (Lyon and Jarvis 1980), and the 2 loci are less than 1 unit apart (Lyon et al. 1981). The circling behavior is caused by widespread abnormalities in the inner ear, comprising severe degeneration of the neural epithelium in the organ of Corti, the maculae, the cristae, and the spiral and vestibular ganglia (Sela et al. 1982). Tissue specimens from the inner ear, maxilla, and tibia of affected males show generalized osteomalacia. Biochemical and histologic evidence of osteomalacia was also reported by Baker et al. (1983) who found, in addition, that affected mice have a renal tubular phosphate leak with normal renal histology. Although the hypophosphatemia and osteomalacia are slightly milder than in animals with the *Hyp* mutation, the gyro mice are more severely affected in several phenotypic features including circling behavior, retarded growth, abnormal face shape, dark coat, male sterility, and tendency to sudden death (Baker et al. 1983). The close proximity to the *Hyp* locus and the biochemical and histologic similarities in the 2 phenotypes suggested that *Gy* may be another hypophosphatemic variant, and Lyon et al. (1986) showed that this was the case. These authors identified a mechanism for hypophosphatemia in the brush-border membrane vesicles; the Na^+/P_i cotransport is approximately 40% that of controls and is similar to that in *Hyp*/Y animals. Lyon et al. (1986) pointed out that this evidence that 2 gene products, coded by different loci on the X chromosome, control renal phosphate reabsorption in the mouse indicates that human X-linked hypophosphatemia is likely to be a heterogeneous phenotype. Subsequently, Boneh et al. (1987) provided evidence suggesting that a form of X-linked hypophosphatemia, in which a sensorineural hearing deficit caused by cochlear dysfunction occurs, is the human counterpart of the gyro phenotype. Vitamin D metabolism appears to be normal in affected mice (Davidai et al. 1988).

Baker LRI, Lyon MF, Goyder Y: 1983. Private communication. *Mouse News Lett* 68: 68 only.

Boneh A, Reade TM, Scriver CR, Rishikof E: 1987. Audiometric evidence for two forms of X-linked hypophosphatemia in humans, apparent counterparts of *Hyp* and *Gy* mutations in mouse. *Am J Med Genet* 27: 997–1003.

Carter TC, Lyon MF, Phillips RJS: 1960. The genetic sensitivity to x-rays of mouse foetal gonads. *Genet Res* 1: 351–355.

Cattanach BM, Pollard CE, Perez JN: 1969. Controlling elements in the mouse X-chromosome. I. Interaction with the X-linked genes. *Genet Res* 14: 223–235.

Davidai GA, Nesbitt T, Drezner MK: 1988. Normal regulation of calcitriol production in *Gy*-mice: evidence for biochemical heterogeneity in the X-linked hypophosphatemic diseases (XLH). (Abstr) *J Bone Miner Res* 3 (Suppl): S92 only.

Lyon MF: 1960. Private communication. *Mouse News Lett* 22: 30 only.

Lyon MF: 1961. Private communication. *Mouse News Lett* 24: 34 only.

Lyon MF: 1966. Private communication. *Mouse News Lett* 35: 28 only.

Lyon MF, Jarvis SE: 1980. Private communication. *Mouse News Lett* 62: 49 only.

Lyon MF, Jarvis SE, Goyder Y: 1981. Private communication. *Mouse News Lett* 64: 56 only.

Lyon MF, Scriver CR, Baker LRI, Tenenhouse HS, Kronick J, Mandla S: 1986. The *Gy* mutation: another cause of X-linked hypophosphatemia in mouse. *Proc Natl Acad Sci USA* 83: 4899–4903.

Sela J, Bab I, Deol MS: 1982. Patterns of matrix vesicle calcification in osteomalacia of Gyro mice. *Metab Bone Dis Relat Res* 4: 129–134.

*30800 HYPOXANTHINE PHOSPHORIBOSYLTRANSFERASE (HPRT; IMP:PYROPHOSPHATE PHOSPHORIBOSYLTRANSFERASE; EC 2.4.2.8)

This is the structural locus for this enzyme which catalyzes one of the first steps in the salvage pathway for the purine bases, hypoxanthine and guanine, in mammalian cells. Bochkarev et al. (1987) provided evidence suggesting a tetrameric structure. HPRT proteins that are functionally identical with each other and with human HPRT have been isolated from hamster (Olsen and Milman 1974), mouse (Hughes et al. 1975), and rat (Gutensohn and Guroff 1972). Mouse, hamster, and human cDNAs exhibit >95% similarity in the coding regions and approximately 80% in the 5′ and 3′ nontranslated regions (Konecki et al. 1982; Chinault and Caskey 1984). Analysis of cDNA recombinants indicate that the human and mouse HPRT genes have 9 exons within 44 kb of genomic DNA; that the intron/exon junctions are identical in both species; and that the 9 exons range in length from 18 to 593 bp in mice and from 18 to 637 bp in humans (Stout and Caskey 1985).

American mink

Rubtsov et al. (1982) used somatic cell hybridization to demonstrate X-linkage.

Cat

O'Brien and Nash (1982) used somatic cell hybridization to demonstrate X-linkage.

Cattle

X-linkage has been demonstrated by Heuertz and Hors-Cayla (1978), Forster et al. (1980), and Shimizu et al. (1981), using cell hybridization.

Chinese hamster

X-linkage was demonstrated by gene dosage studies (Westerveld et al. 1972) and by the use of ovary cells deficient in G6PD and HPRT (Rosenstraus and Chasin 1975). Fuscoe et al. (1983) analyzed 19 hamster cell mutants.

Dasyurids

In these marsupials the locus is syntenic with *GLA, G6PD,* and *PGK* (Dobrovic and Graves 1986); in 2 species, *Sminthopsis crassicaudata* and *Planigales maculata*, the gene order appears to be *G6PD–HPRT–PGK–GLA* with the first-named locus well separated from the others.

Dog

Meera Khan et al. (1984) used somatic cell hybridization to demonstrate X-linkage.

Gibbon

Turleau et al. (1983) used somatic cell hybridization to demonstrate X-linkage.

Horse

Ohno (1973) refers to an unpublished thesis by B. F. Deys in which X-linkage is demonstrated.

Indian muntjac

Shows et al. (1976) demonstrated X-linkage by somatic cell hybridization, and Ropers et al. (1982) assigned the gene to the short arm.

Macropodids (kangaroos and wallabies)

Graves et al. (1979) used marsupial–mouse cell hybrids to demonstrate that the gene for HPRT, together with those for G6PD and PGK, was X-linked, and concluded that the order is *HPRT–PGK–G6PD* in the wallaroo. Although they could not confirm this order, Donald and Hope (1981) used hybrid cell studies to localize the 3 genes to the terminal end of the euchromatic arm of the red kangaroo X chromosome. Subsequently, Dawson and Graves (1984) showed that the gene order was *PGK–HPRT–G6PD* in the gray kangaroo; they did not comment on the difference between this gene order and that proposed by Graves et al.

Mouse, *Hprt*

X-linkage in the mouse was demonstrated by Epstein (1972), who showed that the enzyme level in XO embryos is half that in X embryos at the 2-cell stage. *Mus musculus* and *M. caroli*, a related species from Thailand, have different electrophoretic mobilities for a number of enzymes including HPRT, and Chapman and Shows (1976) and Shows et al. (1976) used cells from fetuses obtained by interspecific hybridization through artificial insemination to confirm X-linkage. Hashimi and Miller (1976) and Francke et al. (1977) confirmed this assignment. Monk and Kathuria (1977) showed that dosage compensation for the enzyme occurs in preimplantation mouse blastocysts. Francke and Taggart (1979, 1980) used a reciprocal X/autosome translocation to show that the gene order in the mouse (centromere–*Hprt–Pgk-1–Ags*) is inverted compared to the order of the homologous loci on the long arm of the human X (centromere–*PGK–GALA–HPRT*). Chapman et al. (1983) used an electrophoretic variation found in 2 species of feral mice that interbreed with laboratory mice to confirm the gene order suggested by Francke and Taggart. Lyon et al. (1987) used hybridization in situ to localize the gene to band A6. Melton et al. (1984) isolated the entire wild-type gene and determined that it is >33 kb long and is split into 9 exons. The HPRT structural variant that occurs in feral mice, which are found in various parts of the world and which represent different taxonomic subgroups of the genus *Mus*, is determined by an allele designated *Hprt a*; the variant occurring in inbred strains is designated *Hprt b* (Chapman et al. 1983). Johnson et al. (1985) showed that murine stocks with the *a* allele have erythrocyte HPRT activity levels approximately 25-fold (*Mus musculus castaneus*) and 70-fold (*Mus spretus*) higher than the levels in laboratory mice with the *b* allele. These differences are due to differences in the turnover rates of the HPRT A and B proteins as reticulocytes mature to erythrocytes. Johnson and Chapman (1987) showed that *a* is common in aboriginal species and several commensal species, whereas *b* is common in feral *M. m. domesticus* populations and inbred strains. Hooper et al. (1987) and Kuehn et al. (1987) produced HPRT⁻ male mice by injecting embryonal stem (ES) cells, which had been selected as HPRT⁻ in tissue culture, into human embryos. The injected cells mingled with the normal embryonic cells, and contributed to all the differentiated tissues, including the germ cells, of the chimeric adults. Gametes derived from the ES cells passed the *Hprt⁻* allele to the next generation. It is not clear how useful the HPRT⁻ mice will be as a model of the Lesch–Nyhan syndrome (hypoxanthine–guanine phosphoribosyltransferase deficiency) (Lesch and Nyhan 1964; Seegmiller et al. 1967); the first mice produced by the transgenic technique appeared to be normal as young adults. The inactive gene in the ES cell line created by Hooper et al. results from a spontaneous deletion (Thompson et al. unpublished results, cited by Doetschman et al. 1987). Doetschman et al. (1987) corrected this defect by reinserting the deleted portion by homologous recombination. Monk et al. (1987) successfully identified *Hprt⁻* male implantation embryos by directly measuring enzyme activity in single cells sampled from individual embryos obtained from female *Hprt⁻* heterozygotes.

Mouse lemur

Cochet et al. (1982) demonstrated X-linkage by using somatic cell hybridization.

Pig

Gellin et al. (1980) demonstrated synteny with *Pgk* and *Hprt*. Forster et al. (1980) and Leong et al. (1983) demonstrated X-linkage by somatic cell hybridization.

Platypus (*Ornithorhynchus anatinus*)

Hprt and *Pgk* are probably syntenic in the platypus (Wrigley and Graves 1987); although it was not possible to assign the loci to a particular chromosome, the pairing is the most ancient known mammalian synteny.

Rabbit

Cianfriglia et al. (1979) and Echard et al. (1981) have shown that this locus is syntenic with other X-linked loci – *Pgk, Gla,* and *G6pd.*

Rat

Yoshida (1978) stated that the locus was "probably" syntenic with *G6pd* and *Pgk*; Levan et al. (1984) demonstrated X-linkage.

Red fox (*Vulpes vulpes*)

Rubtsov et al. (1987) demonstrated X-linkage.

Sheep

Saïdi-Mehtar et al. (1981) provided evidence for synteny with *G6pd, Pgk,* and *Gla.*

Virginia opossum (*Didelphis virginiana*)

The locus is X-linked in this species (Kaslow et al. 1987).

Vole

Cook (1975) demonstrated X-linkage by cell hybridization between mouse cells and vole lymphocytes.

Bochkarev MN, Kulbakina NA, Zhdanova NS, Rubtsov NB, Zakian SM, Serov OL: 1987. Evidence for tetrameric structure of mammalian hypoxanthine phosphoribosyltransferase. *Biochem Genet* 25: 153–160.

Chapman VM, Shows TB: 1976. Somatic cell genetic evidence for X-chromosome linkage of three enzymes in the mouse. *Nature* 159: 665–667.

Chapman VM, Kratzer PG, Quarantillo BA: 1983. Electrophoretic variation for X chromosome-linked hypoxanthine phosphoribosyl transferase (HPRT) in wild-derived mice. *Genetics* 103: 785–795.

Chinault AC, Caskey CT: 1984. The hypoxanthine phosphoribosyltransferase gene: a model for the study of mutation in mammalian cells. *Prog Nucleic Acid Res Mol Biol* 31: 295–313.

Cianfriglia M, Miggiano VC, Meo T, Muller HJ, Muller E, Battistuzzi G: 1979. Evidence for synteny between the rabbit *HPRT, PGK,* and *G6PD* in mouse × rabbit somatic cell hybrids. (Abstr) *Cytogenet Cell Genet* 25: 142 only.

Cochet C, Creau-Goldberg N, Turleau C, de Grouchy J: 1982. Gene mapping of *Microcebus murinus* (Lemuridae): a comparison with man and *Cebus capucinus* (Cebidae). *Cytogenet Cell Genet* 33: 213–221.

Cook PR: 1975. Linkage of the loci for glucose-6-phosphate dehydrogenase and for inosinic acid pyrophosphorylase to the X chromosome of the field-vole *Microtus agrestis. J Cell Sci* 17: 95–112.

Dawson GW, Graves JAM: 1984. Gene mapping in marsupials and monotremes. I. The chromosomes of rodent–marsupial (*Macropus*) cell hybrids, and gene assignments to the X chromosome of the grey kangaroo. *Chromosoma* 91: 20–27.

Dobrovic A, Graves JAM: 1986. Gene mapping in marsupials and monotremes. II. Assignments to the X chromosome of dasyurid marsupials. *Cytogenet Cell Genet* 41: 9–13.

Doetschman T, Gregg RG, Maeda N, Hooper ML, Melton DW, Thompson S, Smithies O: 1987. Targetted correction of a mutant HPRT gene in mouse embryonic stem cells. *Nature* 330: 576–578.

Donald JA, Hope RM: 1981. Mapping a marsupial X chromosome using kangaroo–mouse somatic cell hybrids. *Cytogenet Cell Genet* 29: 127–137.

Echard G, Gellin J, Benne F, Gillois M: 1981. The gene map of the rabbit (*Oryctolagus cuniculus* L.) I. Synteny between the rabbit gene loci coding for HPRT, PGK, G6PD, and GLA: their localization on the X chromosome. *Cytogenet Cell Genet* 29: 176–183.

Epstein CJ: 1972. Expression of the mammalian X chromsome before and after fertilization. *Science* 175: 1467–1468.

Förster M, Stranzinger G, Hellkuhl B: 1980. X-chromosome gene assignment of swine and cattle. *Naturwissenschaften* 67: 48–49.

Francke U, Taggart RT: 1979. Assignment of the gene for cytoplasmic superoxide dismutase (*SOD-1*) to a region of chromosome 16 and of *Hprt* to a region of the X chromosome in the mouse. *Proc Natl Acad Sci USA* 76: 5230–5233.

Francke U, Taggart RT: 1980. Comparative gene mapping: order of loci on the X chromosome is different in mice and humans. *Proc Natl Acad Sci USA* 77: 3595–3599.

Francke U, Lalley PA, Moss W, Ivy J, Minna JD: 1977. Gene mapping in *Mus musculus* by interspecific cell hybridization: assignment of the genes for tripeptidase-1 to chromosome 10, dipeptidase-2 to chromosome 18, acid phosphatase-1 to chromosome 12, and adenylate kinase-1 to chromosome 2. *Cytogenet Cell Genet* 19: 57–84.

Fuscoe JC, Fenwick RG, Ledbetter DH, Caskey CT: 1983. Deletion and amplification of the HGPRT locus in Chinese hamster cells. *Mol Cell Biol* 3: 1086–1096.

Gellin J, Benne F, Hors-Cayla MC, Gillois M: 1980. Carte génique du porc (*Sus scrofa* l.). I. Etude de deux groupes synténiques G6PD, PGK, HPRT et PKM2, MPI. *Ann Génét* 23: 15–21.

Graves JAM, Chew GK, Cooper DW, Johnston PG: 1979. Marsupial–mouse cell hybrids containing fragments of the marsupial X chromosome. *Somatic Cell Genet* 5: 481–489.

Gutensohn W, Guroff G: 1972. Hypoxanthine–guanine–phosphoribosyltransferase from rat brain (purification, kinetic properties, development and distribution). *J Neurochem* 19: 2139–2150.

Hashimi S, Miller OJ: 1976. Further evidence of X-linkage of *hypoxanthine phosphoribosyltransferase* in the mouse. *Cytogenet Cell Genet* 17: 35–41.

Heuertz S, Hors-Cayla M-C: 1978. Carte génétique des bovins par la technique d'hybridation cellulaire. Localisation sur le chromosome de la glucose-6-phosphate déshydrogénase, la phosphoglycérate kinase, l'alpha-galactosidase A et l'hypoxanthine guanine phosphoribosyl transferase. *Ann Génét* 11: 197–202.

Hooper M, Hardy K, Handyside A, Hunter S, Monk M: 1987. HPRT-deficient (Lesch–Nyhan) mouse embryos derived from germ line colonization by cultured cells. *Nature* 326: 292–295.

Hughes SH, Wahl GM, Capecchi MR: 1975. Purification and characterization of mouse hypoxanthine–guanine phosphoribosyltransferase. *J Biol Chem* 250: 120–126.

Johnson GG, Chapman VM: 1987. Altered turnover of hypoxanthine phosphoribosyltransferase in erythroid cells of mice expressing *Hprt a* and *Hprt b* alleles. *Genetics* 116: 313–320.

Johnson GG, Larsen TA, Blakely P, Chapman VM: 1985. Elevated levels of erythrocyte hypoxanthine phosphoribosyltransferase associated with allelic variation of murine *Hprt*. *Biochemistry* 24: 5083–5089.

Kaslow DC, Migeon BR, Persico MG, Zollo M, VandeBerg JL, Samollow PB: 1987. Molecular studies of marsupial X chromosomes reveal limited sequence homology of mammalian X-linked genes. *Genomics* 1: 19–28.

Konecki DS, Brennand J, Fuscoe JC, Caskey CT, Chinault AC: 1982. Hypoxanthine–guanine phosphoribosyltransferase genes of mouse and Chinese hamster: construction and sequence analysis of cDNA recombinants. *Nucleic Acids Res* 10: 6763–6775.

Kuehn MR, Bradley A, Robertson EJ, Evans MJ: 1987. A potential animal model for Lesch–Nyhan syndrome through introduction of HPRT mutations into mice. *Nature* 326: 295–298.

Leong MML, Lin CC, Ruth RF: 1983. The localization of genes for HPRT, G6PD, and alpha-GAL onto the X-chromosome of domestic pig (*Sus scrofa domesticus*). *Can J Genet Cytol* 25: 239–245.

Lesch M, Nyhan WL: 1964. A familial disorder of uric acid metabolism and central nervous system function. *Am J Med* 36: 561–570.

Levan G, Szpirer J, Szpirer C, Hanson C, Islam Q: 1984. The gene map of the rat, *Rattus norvegicus*. (Abstr) *Hereditas* 101: 278 only.

Lyon MF, Zenthon J, Burtenshaw MD, Evans EP: 1987. Localization of the *Hprt* locus by in situ hybridization and distribution of loci on the mouse X chromosome. *Cytogenet Cell Genet* 44: 163–166.

Meera Khan P, Brahe C, Wijnen LMM: 1984. Gene map of dog: six conserved and three disrupted syntenies. (Abstr) *Cytogenet Cell Genet* 37: 537–538.

Melton DW, Konecki Ds, Brennand J, Caskey CT: 1984. Structure, expression, and mutation of the hypoxanthine phosphoribosyltransferase gene. *Proc Natl Acad Sci USA* 81: 2147–2151.

Monk M, Kathuria H: 1977. Dosage compensation for an X-linked gene in pre-implantation mouse embryos. *Nature* 270: 599–601.

Monk M, Hardy K, Handyside A, Whittingham D: 1987. Preimplantation diagnosis of deficiency of hypoxanthine phosphoribosyl transferase in a mouse model for Lesch–Nyhan syndrome. *Lancet* 2: 423–425.

O'Brien SJ, Nash WG: 1982. Genetic mapping in mammals: chromosome map of domestic cat. *Science* 216: 257–265.

Ohno S: 1973. Ancient linkage groups and frozen accidents. *Nature* 244: 259–262.

Olsen AS, Milman G: 1974. Chinese hamster hypoxanthine–guanine phosphoribosyltransferase: purification, structural, and catalytic properties. *J Biol Chem* 249: 4030–4037.

Ropers HH, Sperling K, Raman R, Schmelzer B, Stromaier U, Neitzel H: 1982. Indian muntjac: gene assignment on the short and the long arm of the X chromosome. *Cytogenet Cell Genet* 32: 312 only.

Rosenstraus M, Chasin LA: 1975. Isolation of mammalian cell mutants deficient in glucose-6-phosphate dehydrogenase activity: linkage to hypoxanthine phosphoribosyl transferase. *Proc Natl Acad Sci USA* 72: 493–497.

Rubtsov NB, Radjabli SI, Gradov AA, Serov OL: 1982. Chromosome localization of the genes for isocitrate dehydrogenase-1, isocitrate dehydrogenase-2, glutathione reductase, and phosphoglycerate kinase-1 in the American mink (*Mustela vison*). *Cytogenet Cell Genet* 33: 256–260.

Rubtsov NB, Matveeva VG, Radjabli SI, Kulbakina NA, Nesterova TB, Zakian SM: 1987. Construction of a clone panel of fox–Chinese hamster somatic cell hybrids and assignment of genes for LDHA, LDHB, GPI, ESD, G6PD, HPRT, alpha-GALA in the silver fox. (Russ, Eng summary) *Genetika* 23: 1088–1096.

Saïdi-Mehtar N, Hors-Cayla M-C, Van Cong N: 1981. Sheep gene mapping by somatic cell hybridization: four syntenic groups: ENO1–PGD, ME1–PGM3, LDHB–PEPB–TPI, and G6PD–PGK–GALA. *Cytogenet Cell Genet* 30: 193–204.

Seegmiller JE, Rosenbloom FM, Kelley WN: 1967. Enzyme defect associated with a sex-linked human neurological disorder and excessive purine synthesis. *Science* 155: 1682–1684.

Shimizu N, Shimizu Y, Kondo I, Woods C, Wenger T: 1981. The bovine genes for phosphoglycerate kinase, glucose-6-phosphate dehydrogenase, alpha-galactosidase, and hypoxanthine phosphoribosyltransferase are linked to the X chromosome in cattle–mouse hybrids. *Cytogenet Cell Genet* 29: 26–31.

Shows TB, Brown JA, Chapman VM: 1976. Comparative gene mapping of *HPRT, G6PD,* and *PGK* in man, mouse, and muntjac deer. *Cytogenet Cell Genet* 16: 436–439.

Stout JT, Caskey CT: 1985. HPRT: gene structure, expression, and mutation. *Annu Rev Genet* 19: 127–148.

Turleau C, Creau-Goldberg N, Cochet C, de Grouchy J: 1983. Gene mapping of the gibbon. Its position in primate evolution. *Hum Genet* 64: 65–72.

Westerveld A, Visser RPLS, Freeke MA, Bootsma D: 1972. Evidence for linkage of 3-phosphoglycerate kinase, hypoxanthine–guanine phosphoribosyl transferase, and glucose 6-phosphate dehydrogenase loci in Chinese hamster cells studied by using a relationship between gene multiplicity and enzyme activity. *Biochem Genet* 2: 33–40.

Wrigley JM, Graves JAM: 1987. Conservation of synteny between platypus and man. (Abstr) *Cytogenet Cell Genet* 46: 720 only.

Yoshida MC: 1978. Rat gene mapping by rat–mouse somatic cell hybridization and a comparative Q-banding analysis between rat and mouse chromosomes. *Cytogenet Cell Genet* 22: 606–609.

*30820 ICHTHYOSIS AND MALE HYPOGONADISM

Mouse, scurfy, *sf*

This was the first X-linked mutation discovered in the mouse. It occurred spontaneously in MR-stock at Oak Ridge and was observed in 1949 (Russell et al. 1959). Hemizygous males can be recognized at about 11 days of age by a reddening of the genital papilla. The tail and subsequently other parts of the body are scaly, the skin is "tight," and the eyelids open late. About 2/3 of the affected males die before weaning, and most of the others die shortly afterwards. A few males live for several months but are runty and infertile. Although homozygous females cannot be produced, X^{sf}/O females occurred with a relatively high frequency within a short period after the first affected males were observed. These females exhibited the same phenotypes as hemizygotes (Russell et al. 1959). Heterozygous females are normal (Russell 1964; 733). Lyon (1986) reported that 2- to 3-week-old males appear to be hypogonadal: The testes are very small and located abdominally; the coagulating glands and seminal vesicles are not visible to the naked eye, and there is no scrotum. The absence of the scrotum is often the earliest feature of the disorder; the ichthyosis appears a few days later. Lyon points out that in MIM ichthyosis with hypogonadism (MIM 30820) is listed as a separate disorder from ichthyosis with steroid sulfatase deficiency (MIM 30810). Because scurfy males are not steroid sulfatase deficient (Lyon cites Crocker, personal communication) and sf is located quite separately from Sts, ichthyosis with and without hypogonadism are 2 different entities.

Lyon MF: 1986. Hypogonadism in scurfy (sf) males. *Mouse News Lett* 74: 93 only.
Russell LB: 1964. Another look at the single-active-X hypothesis. *Trans NY Acad Sci Ser II* 26: 726–736.
Russell WL, Russell LB, Gower JS: 1959. Exceptional inheritance of a sex-linked gene in the mouse explained on the basis that the X/O sex chromosome constitution is female. *Proc Natl Acad Sci USA* 45: 554–560.

*30822 IMMUNE RESPONSE TO LDH–C$_4$, TEMPORAL REGULATION OF [NK]

Mouse

Marsh et al. (1977, 1981) demonstrated the existence of a locus involved in regulating the secondary humoral response to LDH–C$_4$. Immediate responsiveness, present in the SJL/J strain, is dominant to delayed responsiveness, present in C3H/HeJ mice. The relation of this locus to *xid* (30030) and others regulating immune responses (30906, 30907, 30908, 31434) is not known.

Marsh JA, Wheat TE, Goldberg E: 1977. Temporal regulation of the immune response to LDH–C$_4$ by an X-linked gene in C3H/HeJ and SJL/J mice. *J Immunol* 118: 2293–2295.
Marsh JA, O'Hern P, Goldberg E: 1981. The role of an X-linked gene in the regulation of secondary humoral response kinetics to sperm-specific LDH–C$_4$ antigen. *J Immunol* 126: 100–106.

30825 IMMUNOGLOBIN M, LEVEL OF

Mouse

There is considerable evidence that X-linked genes determine the serum levels of immunoglobulin M in man. Adinolfi et al. (1978) found evidence for such genes in the mouse: The mean values of IgM were higher in normal females than in males belonging to 2 different strains. However, the well-known relation between IgM levels and the number of X chromosomes in man (46,XY = 45, XO < 46,XX = 47,XXY < 47,XXX) does not exist in the mouse. Hence, the evidence for X-linked genes is meager, and I am surprised that no further studies have been done to confirm and elaborate on the preliminary findings of Adinolfi et al.

Adinolfi M, Haddad SA, Seller MJ: 1978. X chromosome complement and serum levels of IgM in man and mouse. *J Immunogenet* 5: 149–156.

*30838 INTERFERON (IFN) PRODUCTION, EARLY, VIRUS-INDUCED [NK]

Mouse, *If–X*

Zawatsky et al. (1982) described an X-linked locus that influences the levels of serum IFN induced by herpes simplex virus type 1 (HSV-1); a Mendelian analysis carried out in a C57BL/6 × BALB/c cross showed that IFN titers were consistently higher in females than in males. Subsequently, DeMaeyaer-Guignard et al. (1983) showed that an X-linked locus influences early (2–3 h) IFN levels induced by Newcastle disease virus in crosses involving the same 2 strains, which possess different alleles: The IFN levels of male progeny were significantly higher when the X chromosome was of C57BL/6 origin than when it was of BALB/c origin. The sex difference in production is detectable only after puberty, indicating that, in the male, sexual maturation factors influence the expression of the locus. Radiosensitivity studies indicate that the cell populations involved in early IFN synthesis differ from those involved in late IFN synthesis. These findings are also true for the IFN response to HSV-1 (Zawatsky et al. 1983). Mogensen (1977) and Pedersen et al. (1983) demonstrated that natural resistance to herpes simplex virus type 2 (HSV-2) is also an X-linked autointerference phenomenon probably mediated through early IFN production. The results of these experiments conducted in vivo were confirmed in vitro by Ellermann-Ericksen et al. (1986). Because preliminary findings indicate that early production with Sendai virus is also X-linked, DeMaeyer-Guignard et al. (1983) believe that X-linkage of early IFN production may be a general phenomenon.

DeMaeyer-Guignard J, Zawatzky R, Dandoy F, DeMaeyer E: 1983. An X-linked locus influences early serum interferon levels in the mouse. *J Interferon Res* 3: 241–252.

Ellermann-Eriksen S, Liberto MC, Iannello D, Mogensen SC: 1986. X-linkage of the early *in vitro alpha/beta* interferon response of mouse peritoneal macrophages to herpes simplex virus type 2. *J Gen Virol* 67: 1025–1033.

Mogensen SC: 1977. Genetics of macrophage-controlled resistance to hepatitis induced by herpes simplex virus type 2 in mice. *Infect Immun* 17: 268–273.

Pedersen EB, Haarh S, Mogensen SC: 1983. X-linked resistance of mice to high doses of herpes simplex virus type 2 correlates with early interferon production. *Infect Immun* 42: 740–746.

Zawatzky R, Kirchner H, DeMaeyer-Guignard J, DeMaeyer E: 1982. An X-linked locus influences the amount of circulating interferon induced in the mouse by herpes simplex virus type 1. *J Gen Virol* 63: 325–332.

Zawatzky R, Dandoy F, DeMaeyer-Guignard J, DeMaeyer E: 1983. X-linked genes influence early IFN levels in the mouse. In: DeMaeyer E, Schellekens H (eds), *The Biology of the Interferon System 1983.* Amsterdam: Elsevier Science Publishers, pp 77–82.

*30856 IRREGULAR TEETH [?31350]

Mouse, *It*

Although this mutation arose in a radiation experiment, it probably occurred spontaneously. Both lower incisors are markedly reduced or absent in heterozygous females; the upper incisors may be absent also. Expression is variable and may range to normal. Viability and fertility are low. Hemizygotes die in utero (Phipps 1969). Lyon (1974: 258) suggested homology with MIM 31350, TEETH, ABSENCE OF.

Lyon MF: 1974. Mechanisms and evolutionary origins of variable X-chromosome activity in mammals. *Proc R Soc Lond [Biol]* 187: 243–268.

Phipps EL: 1969. Private communication. *Mouse News Lett* 40: 41–42.

30892 LETHAL, X-LINKED, 1 [NK]

Mouse, *sll-1*

Early studies (Morgan 1914: 455; Little 1920) that suggested the existence of a completely recessive sex-linked lethal gene were incomplete in a number of respects (Snell 1931: 56). The evidence presented by Hauschka et al. (1951) for the existence of such a gene in the A strain, although more substantial, was still not entirely convincing (Falconer 1953). More recently, Phillips (1979) described a lethal among the offspring of the daughters of an irradiated male. The affected males, which suckle and appear to be normal, die within about a day of birth.

Falconer DS: 1953. Total sex-linkage in the house mouse. *Z Indukt Abstammungs-Vererbungsl* 85: 210–219.

Hauschka TS, Goodwin MB, Brown E: 1951. Evidence for a sex-linked lethal in the house mouse. *Genetics* 36: 235–253.

Little CC: 1920. Note on the occurrence of a probable sex-linked factor in mammals. *Am Nat* 54: 457–460.

Morgan TH: 1914. Multiple allelomorphs in mice. *Am Nat* 48: 449–458.

Phillips RJS: 1979. Private communication. *Mouse News Lett* 60: 45–46.

Snell GD: 1931. Inheritance in the house mouse: the linkage relations of short-ear, hairless, and naked. *Genetics* 16: 42–74.

30894 LETHAL, X-LINKED, 2 [NK]

Mouse, *sll-2*

See entry 30892 for historical notes. Phillips (1979) found another X-linked lethal in untreated In(X) breeding stocks; it was not described.

Phillips RJS: 1979. Private communication. *Mouse News Lett* 60: 45–46.

*30898 LINED [NK]

Mouse, *Li*

The effects of this mutation, discovered among descendents of an X-irradiated male, were described by Cattanach et al. (1984). Hemizygotes die in utero. In heterozygotes, the coat is finely striped; this is often seen only on parts of the back and can be difficult to detect. The predominance of the wild-type coat appears to be due to nonrandom X chromosome expression; isozyme studies suggest that the *Li* X chromosome is inactive in at least 90% of cells. Heterozygotes appear to be fully viable, although there is a slight shortage of them at weaning age; this deficiency may be caused by prenatal loss or by failure to recognize all *Li*/+ animals. The locus is located distally close to *Hyp* (Cattanach 1985).

Cattanach BM: 1985. Private communication. *Mouse News Lett* 73: 17 only.
Cattanach BM, Crocker AJM, Peters J: 1984. Private communication. *Mouse News Lett* 70: 80–81.

*30906 LYMPHOCYTE ANTIGEN X-1 (LyX-1; IMMUNE RESPONSE TO TYPE III PNEUMOCOCCAL POLYSACCHARIDE) [NK]

Mouse, *(Lyx-1)*

Strains differ in their response to SIII (Amsbaugh et al. 1972), and the trait appears to be regulated by an X-linked locus. The T lymphocytes of high responders bear an antigen, LyX-1, that is not present in low responders (Zeicher et al. 1977). The relation of this locus to *xid* (30030) and others controlling immune responses (30907, 30908, 31434) is not known.

Amsbaugh DF, Hansen CT, Prescott B, Stashak PW, Barthold DR, Baker PJ: 1972. Genetic control of the antibody response to type III pneumococcal polysaccharide in mice. I. Evidence that an X-linked gene plays a decisive role in determining responsiveness. *J Exp Med* 136: 931–948.
Zeicher M, Mozes E, Lonai P: 1977. Lymphocyte alloantigens associated with X-chromosome-linked immune response genes. *Proc Natl Acad Sci USA* 74: 721–724.

*30907 LYMPHOCYTE ANTIGEN X-2 (LyX-2; IMMUNE RESPONSE TO SYNTHETIC DOUBLE-STRANDED RNA) [NK]

Mouse

There are strain differences in response to synthetic poly(I)–poly(C) (Parker and Steinberg 1973). The trait appears to be regulated by an X-linked locus, and the T lymphocytes of high responders bear an antigen, LyX-2, that is not present in low responders (Zeicher et al. 1977). The relation of this locus to *xid* (30030) and others controlling immune responses (30906, 30908, 31434) is not known.

Parker LM, Steinberg AD: 1973. The antibody response to polyinosinic·polycytidylic acid. *J Immunol* 110: 742–751.
Zeicher M, Mozes E, Lonai P: 1977. Lymphocyte alloantigens associated with X-chromosome-linked immune response genes. *Proc Natl Acad Sci USA* 74: 721–724.

*30908 LYMPHOCYTE ANTIGEN X-3 (LyX-3; IMMUNE RESPONSE TO DENATURED DNA) [NK]

Mouse

Strains differ in their ability to respond to denatured DNA (Stoller et al. 1973). Mozes and Fuchs (1974) demonstrated that the trait is regulated by an X-linked gene; high responsiveness is dominant to low responsiveness. T lymphocytes of the former bear an antigen, LyX-3, that is not present in the latter (Zeicher et al. 1977). The relation of this locus to *xid* (30030) and others controlling immune responses (30906, 30907, 31434) is not known.

Mozes E, Fuchs S: 1974. Linkage between immune response potential to DNA and X chromosome. *Nature* 249: 167–168.
Stoller BD, Fuchs S, Mozes E: 1973. Immune response of mice to nucleic acids: strain-dependent differences in magnitude and class of antibody production. *J Immunol* 111: 121–129.
Zeicher M, Mozes E, Lonai P: 1977. Lymphocyte alloantigens associated with X-chromosome-linked immune response genes. *Proc Natl Acad Sci USA* 74: 721–724.

*30909 LYMPHOCYTE-REGULATED, X-LINKED (XLR) [NK]

Mouse

Cohen et al. (1985a) described a T-cell-specific cDNA clone that hybridized with a large X-linked gene family, designated XLR (X-linked, lymphocyte-regulated), which is closely linked to *xid* (30030) at the proximal end of the chromosome. XLR genes are expressed in both B- and T-cell-lineage tumors (Cohen et al. 1985b). This gene family either includes *xid* or is adjacent to it in a gene complex. Siegel et al. (1987) analyzed XLR cDNA clones generated from B-lineage tumors and from thymic tissue, and found a single major XLR transcript expressed by both B and T lymphocytes. Sequence analysis showed that this transcript is capable of coding for a 24,000 mol. wt. protein. The predicted amino acid sequence is strikingly similar to the nuclear envelope constituents, lamins A and C, and to keratin.

Cohen DI, Hedrick SM, Nielsen EA, D'Eustachio P, Ruddle F, Steinberg AD, Paul WE, Davis MM: 1985a. Isolation of a cDNA clone corresponding to an X-linked gene family (XLR) closely linked to the murine immunodeficiency disorder *xid. Nature* 314: 369–372.
Cohen DI, Steinberg AD, Paul WE, Davis MM: 1985b. Expression of an X-linked gene family (XLR) in late-stage B cells and its alteration by the *xid* mutation. *Nature* 314: 372–374.
Siegel JN, Turner CA, Klinman DM, Wilkinson M, Steinberg AD, MacLeod CL, Paul WE, Davis MM, Cohen DI: 1987. Sequence analysis and expression of an X-linked lymphocyte-regulated gene family (XLR). *J Exp Med* 166: 1702–1715.

*30940 MENKES SYNDROME (KINKY-HAIR DISEASE)

Mouse, mottled, *Mo*

Mutations at this locus are common; 9, which have been studied in some detail, are described here. The resultant phenotypes in affected males range from lethal in utero to long-term survival without dietary supplementation. Danks (1986) points out that not only do the mutants share many phenotypic features with the Menkes syndrome but one, blotchy, is very similar to X-linked cutis laxa in man (MIM 30415). He suggests the possibilty "of homology between blotchy and X-linked cutis laxa on the one hand, and between brindled and . . . Menkes' disease on the other hand, with the two sets of loci being closely linked, rather than identical" (p 103). The effects of the *Mo* allele, after which the locus is named, were first described by Fraser et al. (1953). Hemizygotes die at about 11 days of gestation and have no visible external abnormality (Falconer 1953). Heterozygous females have irregular but well-defined patches of full-colored and very lightly colored fur which rarely cross the middorsal or midventral line. The vibrissae are curly, but the coat is not waved. Heterozygotes have reduced viability, but those that survive are usually fertile. Lyon (1960) described a mutation that was probably a recurrence of *Mo*. However, as Silvers (1979) has pointed out, some of the features described by Lyon were not mentioned in the original descriptions of *Mo*. Some affected homozygotes had a tremor that was detectable a few days after birth. Within a few days, the tremor worsened, the animals lost coordination of their limbs and died at about 2 weeks of age. Among those that survived, some developed calcified lumps either attached to the bones or free in the muscles and tendons. The 26K reported by Welshons and Russell (1959) was probably another recurrence of *Mo* (Russell 1960). Silvers (1979: 161–181) reviewed and commented upon the effects of most alleles at the locus.

blotchy, Mo^{blo} This mutation was first described by Russell (1960). Hemizygotes have pale fur and curly whiskers; they are viable and fertile. Heterozygotes have variegated coat color and curly whiskers. The original mutation was assigned a separate locus, but, on the basis of phenotypic similarity, Green (1966) suggested that it was an allele of *Mo*, and this was confirmed by Grahn (1972). Grahn et al. (1969) observed similar macroscopic aortic lesions, often aneurisms preceding hemorrhage, and microscopically abnormal elastic lamellae in homozygous blotchy females, hemizygous blotchy males, and heterozygous dappled (see below) females. Affected males usually live over 5 months, but fail to grow, and they die prematurely from blood vessel rupture. Ranga et al. (1983) developed an outcross line in which Mo^{blo}/Y males breed vigorously. Hemizygotes and homozygotes occasionally have deformed hind limbs. Rowe et al. (1974) described an abnormality of collagen and elastin crosslinking in hemizygotes. This defect, localized to the step at which lysine residues are converted to aldehydes, is expressed as aneurysms of the aorta and its branches, weak skin, and bone deformities. Rowe et al. (1974) postulated that the most likely site of the defect was the enzyme lysyl oxidase or its cofactor, copper. They noted that dopamine beta-hydroxylase is a copper-containing enzyme, drew attention to the error in copper transport in the Menkes syndrome, and suggested that an abnormality in copper metabolism could explain the different defects in mottled mice. Subsequently, Hunt (1974) provided evidence that a primary defect in copper transport underlies the brindled mutant (see below). In Mo^{blo} hemizygote adults, the copper level is markedly increased in the kidney and decreased in the brain (Hunt 1974). Andrews et al. (1975) described the pathogenesis of the aneurysm formation in blotchy mutants: The lesions occur primarily at the points of greatest stress and are confined to the thoracic aorta. The defects of the connective tissue proteins also involve the lung, and emphysema occurs in hemizygotes (Fisk and Kuhn 1976). The outbred Mo^{blo}/Y males developed by Ranga et al. (1983) are structurally normal, but these animals exhibit the abnormalities of connective tissue proteins mentioned previously. Lysyl oxidase activity is markedly reduced in the skin (Rowe et al.1977) and in the lung and cultured fibroblasts of blotchy mice (Starcher et al. 1977). Hemizygotes develop osteoarthrosis of the knee joints from 3 1/2 months of age, and 88% of them are affected when they die approximately 3 months later (Silberberg 1977). Fibroblasts cultured from patients with the Menkes syndrome and from Mo^{blo} hemizygotes have significantly increased levels of copper (Starcher et al. 1978). Mo^{blo}/Y and $Mo^{blo}/+$ animals have a marked alteration of copper transport and a distribution similar to that in brindled mutants of the same age (see below); copper therapy ameliorates many of the effects of the gene but does not improve body weight (Mann et al. 1981). Elevated metallothionein and MT-1 mRNA levels are secondary effects (Packman et al. 1982). Mo^{blo}/Y and $Mo^{blo}/+$ animals lack unique porphyrin- and peroxidase-containing astrocytes from the periventricular areas of the brain (Terr and Weiner 1983).

brindled, Mo^{br}   The effects of this allele were described at the same time as those of *Mo*; however, the 2 mutations arose quite independently (Fraser et al. 1953). Falconer (1956) demonstrated that *Mo* and Mo^{br} are allelic; Grahn (1972) and Lyon (1972) showed that the latter was allelic with Mo^{blo}. Matsushima et al. (1985) described 2 classes of heterozygotes: a "usual type" with a mottled dark and light brown coat, straight whiskers, and no pathologic signs; and a "variant type" with a mottled white and dark-brown coat, curly whiskers, runtiness, tremor, and ataxic gait. In the extreme, these variants resemble hemizygotes but they are fertile and have a normal life-span (see below for additional neurological description). Hemizygous males are virtually devoid of hair pigment, but the eyes are dark and skin pigmentation is normal; the coat is short, the hairs are curved and undulated, and the vibrissae are curly and irregular (Falconer 1953). Grüneberg (1969) presented a detailed description of the variation observed in the hair structure and pigmentation. Affected males show a sustained tremor, a tendency to clasp their hind legs when held up by the tail, a stiffness of the hind limbs, and general inactivity (Falconer 1956). Although the vast majority of the hemizygotes fail to grow, and die at about 14 days, a few survived and were used to produce Mo^{br} females, which have a phenotype identical to the hemizygous males. Unlike mice bearing other Mo alleles, brindled animals do not have aortic lesions (Grahn et al. 1969). Hunt and Johnson (1972a) reported that hemizygotes have a severe reduction in central noradrenaline concentration and an increase in brain tyrosine, and suggested a primary defect in dopamine beta-hydroxylase to explain the former (Hunt and Johnson 1972b). Hunt (1974) provided evidence that a primary defect in copper transport underlies the brindled mutant. He (Hunt 1976) used copper treatment to overcome the lethality, pigment deficiency, and the curly vibrissae in hemizygotes, but the depressed growth rate persisted. Copper levels remained low in liver and brain, and further accumulation occurred in the kidney. The copper-dependent synthesis of brain noradrenaline returned to normal, but the activity of brain cytochrome c oxidase, although increased, remained depressed. Mann et al. (1979a) also observed beneficial effects from a single injection of 50 micrograms of copper at 7 days; however, the initial depression of growth observed by Hunt (1976) was eventually overcome, and, by 60 days, the treated animals had almost normal weights. The activity of lysyl oxidase, which initiates the crosslinking of collagen and elastin, and is copper dependent, is elevated to normal levels in extracts of skin of mice that receive copper therapy (Royce et al. 1982). Copper levels in the brain, liver, and kidney of neonatal heterozygous females approach or equal amounts in untreated hemizygous males; copper levels in the gut are intermediate between those in normals and in hemizygotes. However, in adult heterozygotes, although the kidney accumulation and brain deficiency of copper are maintained, the liver and gut accumulation is considerably reduced (Hunt and Port 1979). The deficiency in the liver and the accumulation in the kidney of hemizygous males are associated with a protein fraction of about 14,000 daltons. Hunt (1976) suggested that the defect may involve an increased affinity for copper in the kidney (and in the gut wall of brindled males) and reduced binding elsewhere. Camakaris et al. (1980) demonstrated abnormalities of copper handling in fibroblasts cultured from the skin of patients with the Menkes syndrome and of brindled mice; Sayed et al. (1981) confirmed the findings in mouse fibroblasts. Danks (1977), in an extensive review of the human and murine homologs, postulated that the basic fault resides in an intracellular copper-binding molecule that is altered so that its copper affinity is increased. Cadmium, nickel, and zinc levels in various tissues from Mo^{br} hemizygotes

are normal (Hunt and Port 1979). Evans and Reis (1978), Camakaris et al. (1979), Port
and Hunt (1979), and Mann et al. (1979a,b) confirmed and extended the observations
of Hunt (1974, 1976). Studies on the kinetics of copper transport and accumulation
by hepatocytes from affected animals demonstrated that hepatic cellular copper
transport per se is normal for both female heterozygotes and male hemizygotes
(Darwish et al. 1983). The only defect observed was the reduced capacity of the
hepatocytes to accumulate copper during continuous incubation in copper-containing
media. These findings suggest the primary inherited defect is expressed in the liver.
On the basis of autoradiographic studies, Delhez et al. (1983) concluded that copper
accumulates in all epithelioid cells in affected animals. The brindled allele – and
probably others – influences tyrosinase and hence melanin pigmentation by acting on
melanocytes to reduce copper levels below those required for a normal complement
of functional tyrosinase (Holstein et al. 1979). Camakaris et al. (1979) and Mann et
al. (1980) found evidence of defective placental transfer of copper; the former group
noted that the copper level in the kidneys of heterozygotes is significantly higher than
that of hemizygotes, and suggested that any differences between the Menkes
syndrome in man and the brindled condition in mice can be explained by known
species differences. In addition, it should be pointed out that in studies on the human
disorder all cases are grouped without any acknowledgment that multiple allelic forms
of the disease may exist in man as they do in the mouse (Procopis et al. 1981).
Hemizygotes have abnormal induction of metallothionein in the kidney (Prins and
Van den Hamer 1980), and they do not display the elevated hepatic metallothionein
synthesis normally observed in 2- to 6-day-old wild-type mice (Piletz and Herschman
1983), but it is unlikely that the mutation directly affects the regulation of
metallothionein production (Brown et al. 1982; Hunt 1983). The metallothionein-I
locus in the mouse (Cox and Palmiter 1983) and in man (Schmidt et al. 1984) is not
located on the X chromosome. The extensive neuronal degeneration of the cerebral
cortex that occurs in hemizygotes (Yajima and Suzuki 1979a) gradually disappears
in the few animals that survive beyond 2 weeks, and cortical neuronal loss and axonal
degeneration of the white matter become the predominant neuropathological features
(Yajima and Suzuki 1979b). These are similar to those in patients with the Menkes
syndrome who survive over 1 year. Cerebellar changes also occur in Mo^{br} hemizygotes between 7 and 31 days but are less evident than the cerebral changes (Nagara
et al. 1980). Abnormal Purkinje cells occur in the cerebellum of heterozygotes; in
young females, these changes are similar to those in young hemizygotes, but in older
females, they are much less conspicuous (Yamano and Suzuki 1986). The "variant
type" heterozygotes described by Matsushima et al. (1985) have ultrastructural
changes that in many respects resemble those in hemizygotes; mild changes occur
rarely in the "usual type" heterozygotes. Nagara et al. (1981) confirmed the findings
of Hunt (1976) and Mann et al. (1979a) on the beneficial clinical effects of copper,
and demonstrated that copper therapy prevents the neuropathological lesions as well;
6–9 months after treatment, there is no recurrence of degenerative changes in the
brains of hemizygotes (Suzuki and Nagara 1981). The dendritic abnormalities in
hemizygotes, which are similar to those in patients with the Menkes syndrome, appear

to result from delayed maturation, and are corrected by treatment with cupric chloride (Yamano and Suzuki 1985). Histopathological features suggestive of copper toxicity occur in the kidneys of treated hemizygotes and heterozygotes. Days 7 and 10 appear to be crucial times in copper homeostasis (Wenk and Suzuki 1982), and treatment on these days increases dopamine beta-hydroxylase in the brains of Mo^{br}/Y mice to normal levels (Wenk and Suzuki 1983). Danks and Camakaris (1983), in an extensive review of the human and mouse disorders, state that the available experimental data could be explained if the primary defect were in the transport of copper, probably at the efflux stage.

dappled, Mo^{dp} This allele was first described by Phillips (1961). Hemizygous males die in utero near term and have thickened and bent ribs, and distorted limb bones and girdles; the cause of death is unknown. In addition to variegated coat color and curly vibrissae, heterozygotes tend to have clubbed hind feet and irregular bony outgrowths from vertebrae, ribs, sternum, and other areas. About 10% of heterozygotes have aortic lesions (Grahn et al. 1969). Studies of chimeras show that Mo^{dp} acts independently in melanocytes and hair follicles (Cattanach et al. 1972).

mosaic, Mo^{ms} This spontaneous mutation was first described by Krzanowska (1966). Although it has not been formally tested for allelism with other mutations at the *Mo* locus, its striking genetic and phenotypic similarity suggests that it is an allele (Green 1981: 164); Silvers (1979: 181) thinks that it may be a recurrence of brindled. Hemizygous males are almost white, and heterozygous females are mottled; the vibrissae in both are curled. Hemizygotes also have altered hair structure (Radochońska 1970). Most affected males do not grow well, develop progressive paresis of the hind limbs, and die in the third postnatal week (Krzanowska 1968; Radochońska 1970). A small percentage of hemizygotes survive and develop well; they have been used to create female homozygotes. Heterozygotes are viable and fertile. When the mutation was placed on an outbred or CBA background, more hemizygotes survived (Wabik 1971); on some inbred backgrounds, some males died in utero. The expression of the allele in their dams also influences the viability of affected males (Styrna 1975, 1977). Mo^{ms} males have impaired copper homeostasis similar to that in brindled mutants (see above) (Styrna 1977).

Neuherberg, Mo^N This mutation arose in the second postirradiation generation after the oocytes of X/O females were irradiated (Schroder 1975). Heterozygotes have irregular patches of fully and lightly colored fur over the whole coat; the vibrissae are curly. The viability of heterozygotes is reduced: About 3% die prenatally and 6–28% die postnatally before weaning; the survivors are fertile and appear normal. Hemizygotes die in utero after they implant. Although no allelism tests were done with other *Mo* alleles, the locus maps close to *Ta* with a recombination frequency similar to that of other alleles. Schroder suggests that this mutation may be a recurrence of *Mo*.

tortoiseshell, Mo^{to}   This mutation was originally described by Dickie (1954). Because it maps close to the *Mo* locus (Lane 1960), interacts with other alleles at the locus (Grahn et al. 1969), and produces a phenotype that is strikingly similar to that produced by these alleles (Rowe et al. 1974), it is considered to be an allele of *Mo* (Green 1981: 164). Mo^{to} is lethal in males, and most die in utero with blood vessel aneurysms; many heterozygotes die of aortic aneurysm rupture (Rowe et al. 1974). These vessel defects result from an abnormality of collagen and elastin crosslinking (see above, Mo^{blo}). Those females that do not succumb to aortic rupture are fertile. Heterozygotes have curly vibrissae and a mottled coat with a slightly silky texture; some females have slight skeletal abnormalities of the hind limbs. Sheedlo and Beck (1981) used an ultrastructural technique to detect copper along the surface of duodenal microvilli and within pinocytotic vesicles at the bases of the microvilli in heterozygotes. The copper deficiency present in heterozygotes causes specific hair anomalies (Sheedlo and Beck 1982).

viable brindled, Mo^{vbr}   Cattanach and Isaacson (1968) described an allele whose effect in hemizygotes is identical with that of brindled except that affected animals survive beyond the 3rd week (Cattanach et al. 1969). However, they are sterile and die between 2 and 3 1/2 months of blood vessel rupture (Grahn et al. 1969). Heterozygotes have a mottled coat that is slightly rippled; the vibrissae are curled. Cattanach and Williams (1972) demonstrated allelism with Mo^{blo}. Affected animals have the same severe reduction in central noradrenaline concentration and increase in brain tyrosine levels observed in Mo^{br}/Y males (Hunt and Johnson 1972a), and the abnormality of collagen and elastin crosslinking described in Mo^{blo} hemizygotes (Rowe et al. 1974). In young (under 3 weeks) hemizygotes, copper levels are markedly increased in kidney and gut, and decreased in liver and brain, but in adult animals only the accumulation in the kidney and the reduction in the brain persists at the same level; in particular, the level in the gut is normal (Hunt and Port 1979).

macular, Mo^{Ml}   This allele occurred as a spontaneous mutation in Japan in 1973. Several reports were published in Japanese, and the studies up to late 1981 were summarized by Nishimura (1982). Subsequent reports on the mutants have been published by Yamano et al. (1987), Kawasaki et al. (1987), Onaga et al. (1987), and Sasahara et al. (1988). Hemizygotes can be distinguished from their littermates by their "white body color" at birth and by their curly whiskers which appear at about 3 days. Seizures and ataxia appear around day 8. Growth stops at about 10 days and then decreases sharply; most animals die at about 16 days, although some live to 22 days. The affected animals are uncoordinated. There are no pathological changes in the cerebral cortex on day 7. However, by day 10, 2 or 3 vacuoles are present in a few neurons in the cerebrum. The number of these neurons gradually increases, and degenerative neurons are present by day 14 (Yamano et al. 1987). The vacuoles are giant mitochondria with an electron-lucent matrix and short peripherally located cristae. A Golgi study (Kawasaki et al. 1987) indicates that dendrite branching is delayed and dendrite arborization is suppressed after day 10. Growth of Purkinje cells is delayed or arrested after day 7. Heterozygotes are not affected neurologically and have a normal life-span. In hemizygotes, copper levels are dramatically increased in the kidney and decreased in liver, brain, and plasma; heterozygotes have abnormal copper metabolism as well. Frequent small (2–5 microgram) injections of $CuCl_2$ effectively correct the neurological disease and increase the life-span to normal. The copper content of the brain and liver in treated hemizygotes is increased significantly; the increase of copper in the cerebral cortex and cerebellum presumably accounts for the improved neurological status of these mice (Katsura et al. 1988). Nishimura (1982) used the treated animals to test for allelism with Mo^{blo} and Mo^{br}.

other mutations   Eicher (1972) described silver-gray, a sex-linked mutant, which resembled blotchy; it probably resulted from another mutation at the *Mo* locus. Fox and Eicher (1978) briefly mentioned another mutation, *pewter* (*pew*), that may be an allele at this locus. Bode et al. (1987) briefly described 2 mutations, induced by ethylnitrosourea, that probably are recurrences of known alleles or new alleles at the locus.

Syrian hamster, mottled, *Mo*

This mutation is lethal in males. Affected females have a thinner than normal coat in which normal-colored fur intermingles with white; they are smaller, less vigorous, and more nervous than their normal littermates (Magalhaes 1954). The mutation was originally assigned the symbol *W*, but this was changed to *Mo*. Ohno (cited by McKusick 1978) maintains that the mutation is homologous to the *Mo* mutation in the mouse and to that causing the Menkes syndrome in man.

Andrews EJ, White WJ, Bullock LP: 1975. Spontaneous aortic aneurysms in *blotchy* mice. *Am J Pathol* 78: 199–210.
Bode VC, Justice MJ, McDonald JD: 1987. Mottled coat color. *Mouse News Lett* 77: 133 only.
Brown R, Camakaris J, Danks DM, Cotton RGH: 1982. Cell hybridization results indicate that the brindled mouse phenotype is recessive. (Abstr) *Am J Hum Genet* 34: 46A only.
Camakaris J, Mann JR, Danks DM: 1979. Copper metabolism in mottled mouse mutants: copper concentrations in tissues during development. *Biochem J* 180: 597–604.
Camakaris J, Danks DM, Ackland L, Cartwright E, Borger P, Cotton RGH: 1980. Altered copper metabolism in cultured cells from human Menkes' syndrome and mottled mouse mutants. *Biochem Genet* 18: 117–131.
Cattanach BM, Isaacson JH: 1968. Private communication. *Mouse News Lett* 38: 17 only.
Cattanach BM, Williams CE: 1972. Private communication. *Mouse News Lett* 47: 34–35.
Cattanach BM, Pollard CE, Perez JN: 1969. Controlling elements in the mouse X-chromosome. I. Interaction with the X-linked genes. *Genet Res* 14: 223–235.
Cattanach BM, Wolfe HG, Lyon MF: 1972. A comparative study of the coats of chimaeric mice and those of heterozygotes for *X*-linked genes. *Genet Res* 19: 213–228.

Cox DR, Palmiter RD: 1983. The metallothionein-I gene maps to mouse chromosome 8: implications for human Menkes' disease. *Hum Genet* 64: 61–64.

Danks DM: 1977. Copper transport and utlisation in Menkes' syndrome and in mottled mice. *Inorg Perspect Biol Med* 1: 73–100.

Danks DM: 1986. Of mice and men, metals and mutations. *J Med Genet* 23: 99–106.

Danks DM, Camakaris J: 1983. Mutations affecting trace elements in humans and animals: a genetic approach to an understanding of trace elements. *Adv Hum Genet* 13: 149–216.

Darwish HM, Hoke JE, Ettinger MJ: 1983. Kinetics of Cu(II) transport and accumulation by hepatocytes from copper-deficient mice and the brindled mouse model of Menkes disease. *J Biol Chem* 258: 13621–13626.

Delhez H, Prins HW, Prinsen L, Van den Hamer CJA: 1983. Autoradiographic demonstration of the copper-accumulating tissues in mice with a defect homologous to Menkes' kinky hair disease. *Pathol Res Pract* 178: 48–50.

Dickie MM: 1954. The tortoiseshell house mouse. *J Hered* 45: 158, 190.

Eicher EM: 1972. Private communication. *Mouse News Lett* 47: 36 only.

Evans GW, Reis BL: 1978. Impaired copper homeostasis in neonatal male and adult female brindled (Mo^{br}) mice. *J Nutr* 108: 554–560.

Falconer DS: 1953. Total sex-linkage in the house mouse. *Z Indukt Abstammungs-Vererbungsl* 85: 210–219.

Falconer DS: 1956. Private communication. *Mouse News Lett* 15: 24 only.

Fisk DE, Kuhn C: 1976. Emphysema-like changes in the lungs of the blotchy mouse. *Am Rev Respir Dis* 113: 787–797.

Fox S, Eicher EM: 1978. Private communication. *Mouse News Lett* 58: 47 only.

Fraser AS, Sobey S, Spicer CC: 1953. Mottled, a sex-modified lethal in the house mouse. *J Genet* 51: 217–221.

Grahn D: 1972. Private communication. *Mouse News Lett* 47: 20 only.

Grahn D, Fry RJM, Hamilton KT: 1969. Genetic and pathologic analysis of the sex-linked allelic series, mottled, in the mouse. (Abstr) *Genetics* 61: S22–S23.

Green MC: 1966. Mutant genes and linkages. In: Green EL (ed), *Biology of the Laboratory Mouse*, 2nd ed. New York: McGraw-Hill, p 90.

Green MC: 1981. Catalog of mutant genes and polymorphic loci. In: Green MC (ed), *Genetic Variants of the Laboratory Mouse*. Stuttgart: Gustav Fisher Verlag, p 164.

Grüneberg H: 1969. Threshold phenomena versus cell heredity in the manifestation of sex-linked genes in mammals. *J Embryol Exp Morphol* 22: 145–179.

Holstein TJ, Fung RQ, Quevedo WC Jr, Bienieki TC: 1979. Effect of altered copper metabolism induced by mottled alleles and diet on mouse tyrosinase. *Proc Soc Exp Biol Med* 162: 264–268.

Hunt DM: 1974. Primary defect in copper transport underlies mottled mutants in the mouse. *Nature* 249: 852–854.

Hunt DM: 1976. A study of copper treatment and tissue copper levels in the murine congenital copper deficiency, mottled. *Life Sci* 19: 1913–1920.

Hunt DM: 1983. Metallothionein and copper homeostasis in mottled mice. (Abstr) *Hereditas* 98: 152 only.

Hunt DM, Johnson DR: 1972a. Aromatic amino acid metabolism in brindled (Mo^{br}) and viable brindled (Mo^{vbr}), two alleles at the mottled locus in the mouse. *Biochem Genet* 6: 31–40.

Hunt DM, Johnson DR: 1972b. An inherited deficiency in noradrenaline biosynthesis in the brindled mouse. *J Neurochem* 19: 2811–2819.

Hunt DM, Port AE: 1979. Trace element binding in the copper deficient mottled mutants in the mouse. *Life Sci* 24: 1453–1466.

Katsura T, Kawasaki H, Yamano T, Shimada M: 1988. Copper contents and pathological changes in various organs of macular mouse. *Congen Anom* 28: 85–92.

Kawasaki H, Onaga A, Yamano T, Shimada M, Nishimura M: 1987. Golgi study on brain of macular mutant mouse as a model of Menkes kinky hair disease. *Acta Neuropathol (Berl)* 72: 349–354.

Krzanowska H: 1966. Private communication. *Mouse News Lett* 35: 35 only.

Krzanowska H: 1968. Private communication. *Mouse News Lett* 38: 25 only.

Lane PW: 1960. Private communication. *Mouse News Lett* 23: 36 only.

Lyon MF: 1960. A further mutation of the mottled type in the house mouse. *J Hered* 51: 116–121.

Lyon MF: 1972. Private communication. *Mouse News Lett* 47: 34 only.

Magalhaes H: 1954. Mottled-white, a sex-linked lethal mutation in the golden hamster, Mesocricetus auratus. (Abstr) *Anat Rec* 120: 752 only.

Mann JR, Camakaris J, Danks DM, Walliczek EG: 1979a. Copper metabolism in mottled mouse mutants: copper therapy of brindled (Mo^{br}) mice. *Biochem J* 180: 605–612.

Mann JR, Camakaris J, Danks DM: 1979b. Copper metabolism in mottled mouse mutants: distribution of ^{64}Cu in brindled (Mo^{br}) mice. *Biochem J* 180: 613–619.

Mann JR, Camakaris J, Danks DM: 1980. Copper metabolism in mottled mouse mutants: defective placental transfer of ^{64}Cu to foetal brindled (Mo^{br}) mice. *Biochem J* 186: 629–631.

Mann JR, Camakaris J, Francis N, Danks DM: 1981. Copper metabolism in mottled mouse (*Mus musculus*) mutants: studies of blotchy (Mo^{blo}) mice and a comparison with brindled (Mo^{br}) mice. *Biochem J* 196: 81–88.

Matsushima H, Okuno A, Eto Y, Maekawa K: 1985. Neuropathologic study in heterozygotes of X-linked brindled mutant mouse. *Acta Neuropathol (Berl)* 67: 300–308.

McKusick VA: 1978. *Mendelian Inheritance in Man: Catalogs of Autosomal Dominant, Autosomal Recessive, and X-Linked Phenotypes*, 5th ed. Baltimore: Johns Hopkins Univ Press, p lxxxix.

Nagara H, Yajima K, Suzuki K: 1980. An ultrastructural study on the cerebellum of the brindled mouse. *Acta Neuropathol (Berl)* 52: 41–50.

Nagara H, Yajima K, Suzuki K: 1981. The effect of copper supplementation on the brindled mouse: a clinico-pathological study. *J Neuropathol Exp Neurol* 40: 428–446.

Nishimura M: 1982. Unusual copper metabolism in the macular mouse. (Japanese) In: Matsushita H (ed), *New Spontaneous Animal Models for Human Disease Developed in Japan*. Tokyo: Seishi Shoin, pp 1–21.

Onaga A, Kawasaki H, Yamano T, Shimada M, Nishimura M: 1987. Light and electron microscopic study of cerebellar cortex of macular mutant mouse as a model of Menkes kinky hair disease. *Brain Dev* 9: 265–269.

Packman S, Palmiter RD, O'Toole C: 1982. Metallothionein mRNA levels in cultured skin fibroblasts of the mottled mouse. (Abstr) *Am J Hum Genet* 34: 59A only.

Phillips RJS: 1961. 'Dappled', a new allele at the *Mottled* locus in the house mouse. *Genet Res* 2: 290–295.

Piletz JE, Herschman HR: 1983. Hepatic metallothionein synthesis in neonatal *mottled–brindled* mutant mice. *Biochem Genet* 21: 465–475.

Port AE, Hunt DM: 1979. A study of the copper-binding proteins in liver and kidney tissue of neonatal normal and mottled mutant mice. *Biochem J* 183: 721–730.

Prins HW, Van den Hamer CJA: 1980. Abnormal copper-thionein synthesis and impaired copper utilization in mutated brindled mice: model for Menkes' disease. *J Nutr* 110: 151–157.

Procopis P, Camakaris J, Danks DM: 1981. A mild form of Menkes steely hair syndrome. *J Pediatr* 98: 97–99.

Radochońska A: 1970. Effect of the gene mosaic (*Ms*) on growth weight, weight of organs and hair structure in mouse. *Genet Polon* 11: 257–274.

Ranga V, Grahn D, Journey TM: 1983. Morphologic and phenotypic analyses of an outcross line of blotchy mouse. *Exp Lung Res* 4: 269–279.

Rowe DW, McGoodwin EB, Martin GR, Sussman MD, Grahn D, Faris B, Franzblau C: 1974. A sex-linked defect in the cross-linking of collagen and elastin associated with the mottled locus in mice. *J Exp Med* 139: 180–192.

Rowe DW, McGoodwin EB, Martin GR, Grahn D: 1977. Decreased lysyl oxidase activity in the aneurysm-prone mottled mouse. *J Biol Chem* 252: 939–942.

Royce PM, Camakaris J, Mann JR, Danks DM: 1982. Copper metabolism in mottled mouse mutants: the effect of copper therapy on lysyl oxidase activity in brindled (Mo^{br}) mice. *Biochem J* 202: 369–371.

Russell LB: 1960. Private communication. *Mouse News Lett* 23: 58–59.

Sasahara A, Yamasaki S, Tachiiri T, Kawasaki H, Yamano T, Ohya N, Shimada M: 1988. Biochemical study on the brain of the macular mutant mouse as a model of Menkes' kinky hair disease. *Brain Dev* 10: 54–56.

Sayed AK, Edwards JA, Bannerman RM: 1981. Copper metabolism of cultured fibroblasts from the brindled mouse (gene symbol Mo^{br}). *Proc Soc Exp Biol Med* 166: 153–156.

Schmidt CJ, Hamer DH, McBride OW: 1984. Chromosomal location of human metallothionein genes: implication for Menkes' disease. *Science* 224: 1104–1106.

Schroder JH: 1975. *Mottled Neuherberg (Mo^N)*, a new male-lethal coat colour mutation of the house mouse (*Mus musculus*). *Theoret Appl Genet* 46: 135–142.

Sheedlo HJ, Beck ML: 1981. Copper localization in the duodenal mucosa of heterozygous tortoiseshell (Mo^{to}/+) female mice (*Mus musculus*). *Genet Res* 38: 333–336.

Sheedlo HJ, Beck ML: 1982. SEM analysis of body hairs and whiskers of heterozygous tortoiseshell (Mo^{to}/+) female mice (*Mus musculus*). *J Anat* 135: 211–216.

Silberberg R: 1977. Epiphyseal growth and osteoarthrosis in blotchy mice. *Exp Cell Biol* 45: 1–8.

Silvers WK: 1979. *The Coat Colors of Mice: A Model for Mammalian Gene Action and Interaction*. New York: Springer-Verlag.

Starcher BC, Madaras JA, Tepper AS: 1977. Lysyl oxidase deficiency in lung and fibroblasts from mice with hereditary emphysema. *Biochem Biophys Res Commun* 78: 706–712.

Starcher B, Madaras JA, Fisk D, Perry EF, Hill CH: 1978. Abnormal cellular copper metabolism in the blotchy mouse. *J Nutr* 108: 1229–1233.

Styrna J: 1975. Survival of *Ms*/– males in two lines of mice selected for a different expression of the gene Mosaic (*Ms*) in heterozygous females. *Genet Polon* 16: 213–219.

Styrna J: 1977. Analysis of causes of lethality in mice with the *Ms* (*Mosaic*) gene. *Genet Polon* 18: 61–79.

Suzuki K, Nagara H: 1981. Brindled mottled mouse: morphological changes of brain and visceral organs in hemizygous males following copper supplementation. *Acta Neuropathol (Berl)* 55: 251–255.

Terr LI, Weiner LP: 1983. Structural and cytochemical changes in astrocytes from the brain periventricular zone of the copper-deficient blotchy mouse. *Anat Rec* 205: 347–353.

Wabik B: 1971. Effect of homo- and heterozygous genetic background on the survival rate of male mice with lethal *mosaic* (*Ms*) gene. *Genet Polon* 12: 545–555.

Welshons WJ, Russell LB: 1959. The Y-chromosome as the bearer of male determining factors in the mouse. *Proc Natl Acad Sci USA* 45: 560–566.

Wenk G, Suzuki K: 1982. The effect of copper supplementation on the concentration of copper in the brain of the brindled mouse. *Biochem J* 205: 485–487.

Wenk G, Suzuki K: 1983. Congenital copper deficiency: copper therapy and dopamine-beta-hydroxylase activity in the mottled (brindled) mouse. *J Neurochem* 41: 1648–1652.

Yajima K, Suzuki K: 1979a. Neuronal degeneration in the brain of the brindled mouse – a light microscope study. *J Neuropathol Exp Neurol* 38: 35–46.

Yajima K, Suzuki K: 1979b. Neuronal degeneration in the brain of the brindled mouse. I. Chronological studies on the long-surviving group. *Acta Neuropathol (Berl)* 48: 127–132.

Yamano T, Suzuki K: 1985. Abnormalities of Purkinje cell arborization in brindled mouse cerebellum: a Golgi study. *J Neuropathol Exp Neurol* 44: 85–96.

Yamano T, Suzuki K: 1986. Cerebellar changes of the female mice heterozygous for brindled gene. *Acta Neuropathol (Berl)* 69: 220–226.

Yamano T, Shimada M, Kawasaki H, Onaga A, Nishimura M: 1987. Clinico-pathological study on macular mutant mouse. *Acta Neuropathol (Berl)* 72: 256–260.

*30986 MOLONEY MURINE LEUKEMIA VIRUS INTEGRATION SITE-14 (M-MuLV INTEGRATION SITE-14) [NK]

Mouse, *Mov-14*

By infecting mouse embryos with Moloney murine leukemia virus (M-MuLV), Jaenisch et al. (1981) produced several mouse substrains with stable germ line integration of retroviral DNA at distinct chromosomal loci (*Mov* loci). One of these integration sites (*Mov-14*) is on the X chromosome (Stewart et al. 1983).

Jaenisch R, Jahner D, Nobis P, Simon I, Lohler J, Harbers K, Grotkopp D: 1981. Chromosomal position and activation of retroviral genomes inserted into the germ line of mice. *Cell* 24: 519–529.

Stewart C, Harbers K, Jähner D, Jaenisch R: 1983. X chromosome-linked transmission and expression of retroviral genomes microinjected into mouse zygotes. *Science* 221: 760–762.

*30987 MOLONEY MURINE LEUKEMIA VIRUS INTEGRATION SITE-15 (M-MuLV INTEGRATION SITE-15)) [NK]

Mouse, *Mov–15*

Münke et al. (1986) noted in a table that this integration site is located in the pairing region and is also on the Y chromosome.

Münke M, Harbers K, Jaenisch R, Franke U: 1986. Chromosomal mapping of four different integration sites of Moloney murine leukemia virus including the locus for alpha 1 (I) collagen in mouse. *Cytogenet Cell Genet* 43: 140–149.

*31010 MUSCULAR DYSTROPHY, DUCHENNE AND BECKER TYPES (DMD)

Dog, canine-X-linked muscular dystrophy (CXMD)

Wentink et al. (1972) reported on a primary myopathy in 5 male Irish terrier pups. Although no bitch-related animals were available for clinical investigation, the pedigree indicated X-linked inheritance. The affected dogs developed normally until 8 weeks, at which time they had trouble swallowing; by 13 weeks, they had difficulty walking. An enlarged tongue and progressive muscular atrophy were noted. Neurological signs were normal. Almost all blood enzymes investigated were markedly elevated; creatine phosphokinase and aldase were extremely high. At necropsy, the muscles were pale with yellowish white streaks. Histologically, the distribution of the lesions was patchy, and there were granular and floccular changes with phagocytosis, giant cells, and calcifications. Although histochemical changes were the same in all muscles studied, they varied in severity. Abnormal mitochondria and unidentified electron-dense bodies were observed ultrastructurally. Lyon (1974: 258) suggested that this disorder may be homologous with the human Emery–Dreifuss type of muscular dystrophy (MIM 31030). I do not know if this mutation still exists. A similar disorder has been recognized in male golden retriever dogs (Meier 1958; Cardinet and Holliday 1979; deLahunta 1983; Kornegay 1986; Valentine et al. 1986). Valentine et al. (1986) noted that, although some differences exist, the features are strikingly similar to those in human Duchenne muscular dystrophy; these observations were extended by Cooper et al. (1988a). The occurrence of the disorder only in males suggested X-linked inheritance, and this was confirmed by appropriate crosses (Cooper et al. 1988b). Cooper et al. (1988a) showed that affected dogs lack the DMD gene transcript and its product, dystrophin.

Mouse, X-linked muscular dystrophy (*mdx*)

This mutation, which arose spontaneously, results in a 3- to 5-fold increase in normal levels of blood pyruvate kinase (PK), and was originally called pyruvate kinase expression (*pke*) (Moore and Bulfield 1981). However, subsequent studies of the mutants revealed that the extra PK activity resides in the plasma, which normally has almost no PK activity; that the plasma contains large amounts of the muscle type of creatinine phosphokinase isozyme; and that the skeletal muscles are dystrophic. Consequently, the mutation was renamed X-linked muscular dystrophy (*mdx*) (Bulfield et al. 1982). Bulfield et al. (1984) described the details of the discovery of the mutant, the nature of the myopathic lesions, and linkage relations. Affected heterozygotes and homozygotes have mild clinical signs and are viable and fertile. The central nervous system appears normal. Heterozygotes appear normal and show no histological or biochemical changes. Several histological and ultrastructural investigations of the pathogenetic changes in skeletal muscle from hemizygous males and homozygous females have been reported (Dangain and Vrbova 1984; Bridges 1986; Tanabe et al. 1986; Anderson et al. 1987; Carnworth and Shotton 1987; Torres and Duchen 1987). During the first few weeks of postnatal life, degeneration of muscle fibers can be detected; this is followed by regeneration. The degeneration/re-generation cycle, which also occurs in cardiac muscle (Bridges 1986), continues throughout the life of the animals, although the process tends to be slower in adults. This picture is similar to that observed in the muscles of boys with DMD (Cullen and Mastaglia 1980), the most common form of human X-linked dystrophy. However, in the human disorder, the degeneration/regeneration process is accompanied by adipose and endomysial connective tissue infiltration that gradually replaces much of the muscle mass and ultimately leads to death when the respiratory muscles fail. This critical connective tissue infiltration is absent in the murine disorder. Hoffman et al. (1987a) isolated and characterized cDNAs corresponding to the mouse equivalent of the human DMD gene, and found that the gene is expressed equally well in the skeletal and cardiac tissue of both species. The nucleic acid and predicted amino acid sequences are nearly 90% congruent in the 2 species. The *mdx* locus is situated in the proximal half of the chromosome (Bulfield and Isaacson 1985; Chapman et al. 1985). Brockdorff et al. (1987) used an interspecific mouse *domesticus/spretus* cross to position sequences homologous to a DMD cDNA clone on the mouse X chromosome "provocatively close" to *mdx*. The DMD sequences are distant from *spf*, the homolog of *OTC* in man. Because *OTC* and *DMD* are closely linked, Brockdorff et al. state that one of the major ancestral breakpoints that have been postulated to explain the evolution of the mouse and human X chromosomes (see Introduction, Fig. 1) is located between these 2 loci. Heilig et al. (1987), using 2 sequences from *DMD* that cross-hybridize to mouse X-linked sequences, localized the homologous mouse gene. Both sequences are in a region of 10 centimorgans between the tabby (*Ta*) and *DXPas8* (see Introduction, Appendix I) loci close to the phosphorylase kinase 6 (*Phk*) locus. These authors conclude that the region of the mouse X chromosome around the glucose-6-phosphate dehydrogenase (*G6pd*) and *DXPas8* loci contains 2 genes that give rise to 2 distinct human disorders: Emery–Dreifuss muscular dystrophy, the gene for which is known to be closely linked to *DXS52* (the homolog of *DXPas8*), and DMD. Chamberlain et al. (1987) also mapped the mouse *DMD* gene close to the *mdx* locus. Hoffman et al. (1987b) used polyclonal antibodies directed against fusion proteins containing 2 distinct regions of mDMD cDNA to identify the protein product of *DMD* and *mdx*. The protein, dystrophin, is approximately 400 kd and represents about 0.002% of total striated muscle protein. Muscle isolated from affected boys and mice contains no detectable dystrophin. The striking difference in the typical consequence of this deficiency in the 2 species might be due to differences in secondary biochemical effects or histological changes. Bonilla et al. (1988)

immunohistochemically localized dystrophin to the sarcolemma of normal human and mouse skeletal fibers. The protein is absent from muscle from mdx mice and absent or markedly deficient in muscle from patients with DMD. Mouse *Dmd* mRNA is present in mouse skeletal and cardiac muscle, and in brain (Chamberlain et al. 1988). It is also present at much lower levels in the muscle and brain of 3 different strains of dystrophic (*mdx* variants) mice. These results provide additional evidence that *mdx* mutations are alleles of *Dmd*; however, the data of Grant et al. (1988) indicate that *mdx* and *Dmd* are separate loci. Nudel et al. (1988) found that the putative *DMD* gene is developmentally regulated in rat and mouse myogenic cell cultures, and is expressed in rat and mouse striated muscle, in mouse smooth muscle, and in rat, mouse, and rabbit brain.

Rat

Nudel et al. (1988) studied the regulation and expression of the putative *DMD* gene in tissues of the rat (see under Mouse for details).

Anderson, JE, Ovalle WK, Bressler BH: 1987. Electron microscopic and autoradiographic characterization of hindlimb muscle regeneration in the mdx mouse. *Anat Rec* 219: 243–257.

Bonilla E, Samitt CE, Miranda AF, Hays AP, Salviati G, Mauro S, Kunkel LM, Hoffman EP, Rowland LP: 1988. Duchenne muscular dystrophy: deficiency of dystrophin at the muscle cell surface. *Cell* 54: 447–452.

Bridges LR: 1986. The association of cardiac muscle necrosis and inflammation with the degenerative and persistent myopathy of *mdx* mice. *J. Neurol Sci* 72: 147–157.

Brockdorff N, Cross GS, Cavanna JS, Fisher EMC, Lyon MF, Davies KE, Brown SDM: 1987. The mapping of a cDNA from human X-linked Duchenne muscular dystrophy gene to the mouse X chromosome. *Nature* 328: 166–168.

Bulfield G, Isaacson JH: 1985. Private communication. *Mouse News Lett* 72: 97 only.

Bulfield G, Siller WG, Wight PAL: 1982. Private communication. *Mouse News Lett* 66: 57 only.

Bulfield G, Siller WG, Wight PAL, Moore KJ: 1984. X chromosome-linked muscular dystrophy (*mdx*) in the mouse. *Proc Natl Acad Sci USA* 81: 1189–1192.

Cardinet GH III, Holliday TA: 1979. Neuromuscular diseases of domestic animals: a summary of muscle biopsies from 159 cases. *Ann NY Acad Sci* 317: 290–311.

Carnworth JW, Shotton DM: 1987. Muscular dystrophy in the *mdx* mouse: histopathology of the soleus and extensor digitorum longus muscles. *J Neurol Sci* 80: 39–54.

Chamberlain JS, Grant SG, Reeves AA, Mullins LJ, Stephenson DA, Hoffman EP, Monaco AP, Kunkel LM, Caskey CT, Chapman VM: 1987. Regional localization of the murine Duchenne muscular dystrophy gene on the mouse X chromosome. *Somatic Cell Mol Genet* 13: 671–678.

Chamberlain JS, Pearlman JA, Muzny DM, Gibbs RA, Ranier JE, Reeves AA, Caskey CT: 1988. Expression of the murine Duchenne muscular dystrophy gene in muscle and brain. *Science* 239: 1416–1418.

Chapman V, Murawski M, Miller D, Swiatek D: 1985. Private communication. *Mouse News Lett* 72: 120 only.

Cooper BJ, Winand NJ, Steadman H, Valentine BA, Hoffman EP, Kunkel LM, Scott M-O,

Fischbeck KH, Kornegay JN, Avery RJ, Williams JR, Schmickel RD, Sylvester JE: 1988a. The homologue of the Duchenne locus is defective in X-linked muscular dystrophy of dogs. *Nature* 334: 154–156.

Cooper BJ, Valentine BA, Wilson S, Patterson DF, Concannon PW: 1988b. Canine muscular dystrophy: confirmation of X-linked inheritance. *J Hered* 79: 405–408.

Cullen MJ, Mastaglia FL: 1980. Morphological changes in dystrophic muscle. *Br Med Bull* 36: 145–152.

Dangain J, Vrbova G: 1984. Muscle development in mdx mutant mice. *Muscle Nerve* 7: 700–704.

deLahunta A: 1983. *Veterinary Neuroanatomy and Clinical Neurology*, 2nd ed. Philadelphia: W. B. Saunders, pp 84–88.

Grant SG, Mullins LJ, Stephenson DA, Chapman VM: 1988. Mapping of *mdx* and *Dmd. Mouse News Lett* 80: 181 only.

Heilig R, Lemaire C, Mandel J-L, Dandolo L, Amar L, Avner p: 1987. Localization of the region homologous to the Duchenne muscular dystrophy locus on the mouse X chromosome. *Nature* 328: 168–170.

Hoffman EP, Monaco AP, Feener CC, Kunkel LM: 1987a. Conservation of the Duchenne muscular dystrophy gene in mice and humans. *Science* 238: 347–350.

Hoffman EP, Brown RH Jr, Kunkel LM: 1987b. Dystrophin: the protein product of the Duchenne muscular dystrophy locus. *Cell* 51: 919–928.

Kornegay JN: 1986. Golden retriever myopathy. In: Kirk RW (ed), *Current Veterinary Therapy, IX*. Philadelphia: W. B. Saunders, pp 792–793.

Lyon MF: 1974. Mechanisms and evolutionary origins of variable X-chromosome activity in mammals. *Proc R Soc Lond [Biol]* 187: 243–268.

Meier H: 1958. Myopathies in the dog. *Cornell Vet* 48: 313–330.

Moore K, Bulfield G: 1981. Private communication. *Mouse News Lett* 64: 61 only.

Nudel U, Robzyk K, Yaffe D: 1988. Expression of the putative Duchenne muscular dystrophy gene in differentiated myogenic cell cultures and in the brain. *Nature* 331: 635–638.

Tanabe Y, Esaki K, Nomura T: 1986. Skeletal muscle pathology in X chromosome-linked muscular dystrophy (*mdx*) mouse. *Acta Neuropathol (Berl)* 69: 91–95.

Torres LFB, Duchen LW: 1987. The mutant *mdx*: inherited myopathy in the mouse morphological studies of nerves, muscles and end-plates. *Brain* 110: 269–299.

Valentine BA, Cooper BJ, Cummings JF, deLahunta A: 1986. Progressive muscular dystrophy in a golden retriever dog: light microscope and ultrastructural features at 4 and 8 months. *Acta Neuropathol (Berl)* 71: 301–310.

Wentink GH, van der Linde-Sipman JS, Meijer AEFH, Kamphuisen HAC, van Vorstenbosch CHAHV, Hartman W, Hendriks HJ: 1972. Myopathy with a possible recessive X-linked inheritance in a litter of Irish terriers. *Vet Pathol* 9: 328–349.

*31047 NADH–COENZYME Q REDUCTASE DEFICIENCY (ELECTRON TRANSPORT CHAIN, DEFECT OF COMPLEX I OF) [NK]

Chinese hamster

Scheffler and his colleagues isolated and characterized several different respiration-deficient mutants of Chinese hamster cells in culture (DeFrancesco et al. 1976; Ditta et al. 1976; Breen and Scheffler 1979; Soderberg et al. 1979). They can be sorted into 7 complementation groups, 5 of which appear to be defective in complex I of the electron transport chain. In the inner mitochondrial membrane, this membrane-bound complex (also known as NADH–coenzyme Q reductase) contains at least 25 different polypeptides. Day and Scheffler (1982) showed that 3 mutations associated with 2 different complementation groups map to the X chromosome.

Mouse

The 3 mutations that are X-linked in the Chinese hamster (above) are also X-linked in the mouse (Day and Scheffler 1982).

Breen GAM, Scheffler IE: 1979. Respiration-deficient Chinese hamster cell mutants: biochemical characterization. *Somatic Cell Genet* 5:441–451.
Day CE, Scheffler IE: 1982. Mapping of the genes of some components of the electron transport chain (complex I) on the X chromosome of mammals. *Somatic Cell Genet* 8: 691–707.
DeFrancesco L, Scheffler IE, Bissell MJ: 1976. A respiration-deficient Chinese hamster cell line with a defect in NADH–coenzyme Q reductase. *J Biol Chem* 251:4588-4595.
Ditta G, Soderberg K, Landy F, Scheffler IE: 1976. The selection of Chinese hamster cells deficient in oxidative energy metabolism. *Somatic Cell Genet* 2:331–344.
Soderberg K, Mascarello JT, Breen GAM, Scheffler IE: 1979. Respiration-deficient Chinese hamster cell mutants: genetic characterization. *Somatic Cell Genet* 5:225–240.

*31048 NEPHRITIS, HEREDITARY, X-LINKED (SAMOYED HEREDITARY GLOMERULOPATHY; SHG) [?30105]

Dog

Bernard and Valli (1977) described a form of hereditary nephritis in a family of Canadian Samoyed dogs; breeding data, updated by Bloedow (1981) and Jansen et al. (1986a), suggest X-linkage. Renal disease is first detected between 2 and 3 months of age with the onset of proteinuria, microscopic hematuria, and hypoalbuminemia. Additional features of renal failure become manifest, and affected animals die between 8 and 15 months. Jansen et al. (1987) described the course of the disease in 11 males. Lesions of the glomerular capillary basement membranes (GCBM) are detectable by electron microscopy as early as 1 month (Jansen et al. 1984, 1986b). Discrete portions of the GCBM show subepithelial and subendothelial lucency, and reduplication or splitting of the lamina densa into multiple layers, with small round electron-dense particles between the layers; many capillary loops appear normal. The lesions gradually become more severe and widespread, and eventually all glomerular capillaries are affected. Glomerulosclerosis is complete by 8–10 months. The absence of the Goodpasture antigen (GPA) from the GCBM of affected males correlates with their multilaminar splitting (Thorner et al. 1987). The results of immunofluorescent and electron microscopic studies suggest that the GPA is not required to form trilaminar GCBM, but is necessary subsequently to maintain their integrity. The GPA is normally present in the C-terminal (MC1) domain of the collagen type IV molecule, and Thorner et al. (1987) postulate that the human and canine disorders result from a defect in this domain. The ultrastructural changes in the GCBM are strikingly similar to those in the human disorder. Thorner et al. (1988) studied the basement membranes of several extrarenal tissues including the lens, retina, and inner ear, and did not detect any differences between affected and nonaffected dogs, except that Goodpasture antigen (GPA) was not detected in the lens capsule, the internal limiting membrane of the retina, and the basilar membrane of the inner ear of affected dogs. In carrier females, the eye tissues possessed the antigen. Affected dogs appear to have normal vision and hearing. Heterozygous females show proteinuria and splitting of the GCBM that is considerably less severe than in affected males (Jansen et al. 1986a,b). Among the females, 1 proven carrier and 1 "normal" died of the kidney disorder; they showed clinical signs similar to affected males only during the last 8 weeks of life. The carrier died at 8 years and 9 months, and the "normal" at 5 years. Two other carriers are reported to have had a mild proteinuria for 2 years (Bloedow 1981). Jansen et al. (1986a) do not mention these animals, although the pedigree they present contains 3 bitches that may be severely affected. Jansen et al. (1987) described the course of the disorder in 4 carrier females. Although the segregation pattern appears to be reasonably compatible with X-linkage, there are a number of unsatisfactory aspects, notably an overall lack of female pups (sex ratio, 1.3:1; N = 108), particularly carriers: Of 37 bitches born to 7 carrier females, only 8 were proven heterozygotes. The Alport syndrome (nephritis–deafness syndrome) in man is genetically hetero-geneous; 1 form is X-linked (MIM 30105). Until more reliable data become available, the relation of this disorder in the dog to the human condition is conjectural. A colony of these dogs is maintained at the Ontario Veterinary College, Guelph, Ontario, Canada.

Bernard MA, Valli VE: 1977. Familial renal disease in Samoyed dogs. *Can Vet J* 18: 181–189.

Bloedow AG: 1981. Familial renal disease in Samoyed dogs. *Vet Rec* 108: 167–168.

Jansen B, Thorner PS, Singh A, Patterson JM, Lumsden JH, Valli VE, Baumal R, Basrur PK: 1984. Hereditary nephritis in Samoyed dogs. *Am J Pathol* 116: 175–178.

Jansen B, Tryphonas L, Wong J, Thorner P, Maxie MG, Valli VE, Baumal R, Basrur PK: 1986a. Mode of inheritance of Samoyed hereditary glomerulopathy: an animal model of hereditary nephritis in humans. *J Lab Clin Med* 107: 551–555.

Jansen B, Thorner P, Baumal R, Valli V, Maxie MG, Singh A: 1986b. Samoyed hereditary glomerulopathy (SHG): evolution of splitting of glomerular capillary basement mem-branes. *Am J Pathol* 125: 536–545.

Jansen B, Valli VEO, Thorner P, Baumal R, Lumsden JH: 1987. Samoyed hereditary glomerulopathy: serial, clinical and laboratory (urine, serum biochemistry and hematol-ogy) studies. *Can J Vet Res* 51: 387–393.

Thorner P, Jansen B, Baumal R, Valli VE, Goldberger A: 1987. Samoyed hereditary glomerulopathy: immunohistochemical staining of basement membranes of kidney for laminin, collagen type IV, fibronectin, and Goodpasture antigen, and correlation with electron microscopy of glomerular capillary basement membranes. *Lab Invest* 56: 435–443.

Thorner PS, Jansen B, Baumal R, Harrison RV, Mount RJ, Valli VEO, Spicer PM, Marrano PM: 1988. An immunohistochemical and electron microscopic study of extra-renal basement membranes in dogs with Samoyed hereditary glomerulopathy. *Virchows Arch[A]* 412: 281–290.

*31123 ORANGE [NK]

Cat, *O*

This mutation, which produces the ginger or marmalade coat color, was originally called yellow, but the standard designation is now orange (Robinson 1977). Hemizygous males and homozygous females are orange-yellow, and heterozygous females are mosaics of wild type and orange-yellow (the tortoiseshell phenotype). Little (1912) and Doncaster (1912) concluded independently that the gene is X-linked, and Doncaster (1913) published a large body of breeding data supporting this view. The occasional occurrence of apparent exceptions, notably tortoiseshell males and "black" (non orange) females, was a source of controversy for 50 years until the development of mammalian chromosome techniques (Thuline and Norby 1961) and new knowledge of X chromosome activity (Lyon 1962) led to a resolution. However, a large proportion of tortoiseshell males are 38XY and are fertile; the evidence suggests that they arise from somatic reversion (gene instability) (Moran et al. 1984). C. C. Little suggested this possibility in 1912. The action of the gene converts eumelanin (black or brown pigment) to pheomelanin (orange pigment) in the melanocytes of the hair follicles. The interaction of the *O* gene with others to produce various coat color phenotypes is described by Robinson (1977). This gene is probably homologous with *To* (*tortoiseshell*) in the Syrian hamster (see below); however, according to Searle (1968), *O* affects skin as well as hair pigment, whereas *To* affects only the latter.

Syrian (golden) hamster, tortoiseshell, *To*

Hemizygous males and homozygous females are orange-yellow. The gene converts eumelanin to pheomelanin in the melanocytes of the hair follicles. Eumelanin is eliminated in the hair only; normal amounts of dark pigment develop in the skin. Heterozygous females are a mosaic of yellow and agouti (hence, tortoiseshell) (Robinson, 1966). This mutation is probably homologous with *O* (orange) in the cat (see above).

Doncaster L: 1912. Sex-limited inheritance in cats. *Science* 36: 144 only.

Doncaster L: 1913. On sex-linked inheritance in cats, and its bearing on the sex-limited transmission of certain human abnormalities. *J Genet* 3: 11–23.

Little CC: 1912. Preliminary note on the occurrence of a sex-limited character in cats. *Science* 35: 784–785.

Lyon MF: 1962. Sex chromatin and gene action in the mammalian X-chromosome. *Am J Hum Genet* 14: 135–148.

Moran C, Gillies CB, Nicholas FW: 1984. Fertile male tortoiseshell cats: mosaicism due to gene instability? *J Hered* 75: 397–402.

Robinson R: 1966. Sex-linked yellow in the Syrian hamster. *Nature* 212: 824–825.

Robinson R: 1977. *Genetics for Cat Breeders*, 2nd ed. Oxford: Pergamon Press, pp 112–118.

Searle AG: 1968. *Comparative Genetics of Coat Color in Mammals*. London: Logos Press, p 110.

Thuline HC, Norby DE: 1961. Spontaneous occurrence of chromosome abnormality in cats. *Science* 134: 554–555.

*31124 ORNITHINE DECARBOXYLASE, X-LINKED (ODC; L-ORNITHINE CARBOXY-LYASE; EC 4.1.1.17 [NK]

Mouse, *Odc-13*

This is the first enzyme in polyamine biosynthesis (Pegg and McCann 1982). In the mouse, *Odc* is a dispersed gene family; one member is on the X chromosome in the region between *G6pd* and *Ags* (Elliott et al. 1988; Stephenson et al. 1988).

Elliott RW, Stephenson DA, Grant SG, Chapman VM: 1988. Identification of an X-linked member of the *Odc* gene family. *Mouse News Lett* 80: 180–181.

Pegg AE, McCann PP: 1982. Polyamide metabolism and function. *Am J Physiol* 243: C212–C221.

Stephenson DA, Elliott RW, Chapman VM, Grant SG: 1988. Identification of an X-linked member of the *Odc* family in the mouse. *Nucleic Acids Res* 16: 1642 only.

*31125 ORNITHINE TRANSCARBAMYLASE (ORNITHINE CARBAMOYLTRANSFERASE; OTC; EC 2.1.3.3)

This is the structural locus for this homotrimeric, mitochondrial matrix enzyme of the urea acid cycle in mammals; it is expressed in the liver and, to a lesser degree, in the intestinal mucosa and visceral yolk sac endoderm (Grant et al. 1987).

Marsupials and monotremes

The OTC locus is on an autosome in distantly related marsupial species (*Dasykaluta rosamondae* and *Macropus rufus*) and a monotreme species (the platypus, *Ornithorhynchus anatinus*) (Sinclair et al. 1987). This is the first demonstration that an X-linked gene in one mammalian species is autosomally linked in another, although the evidence that the marsupial STS locus is also autosomal is compelling (31341). Sinclair et al. propose that, in a common evolutionary ancestor of the 3 mammalian groups, the OTC locus was in autosomal or pseudoautosomal region and was not subjected to X-inactivation, and that recently in eutherian evolution it was translocated or incorporated into the inactivated region of the X chromosome, whereas, in metatherian and protherian mammals, it remained autosomal or was translocated to an autosome. The American marsupial, the Virginia opossum, has no highly conserved homolog to human *OTC* cDNA (Kaslow et al. 1987).

Mouse

Scherer et al. (1988) isolated and characterized the mouse OTC gene, which spans about 70 kb and is composed of 10 exons. The protein coding regions of the mouse, rat, and human cDNA clones are each 1,062 bases long, and the murine gene is 96% and 88% congruent with the rat and human genes in this region.

sparse fur, spf   The effects of this mutation were first described by Cupp (1958) and Russell (1960). It appears to be homologous with a mutation causing a deficiency in man that often produces lethal intoxication due to hyperammonemia in affected males. The first mutant was observed in a descendant, several generations removed, of an irradiated male. Affected animals (hemizygous males and homozygous females) are small, with wrinkled skin and little or no fur. On some backgrounds, *spf*/Y males may become fully furred and attain normal size by the age of weaning. Most heterozygotes appear normal, but a few have the mutant phenotype. *spf*/Y males have a high mortality rate. Affected animals frequently have urinary bladder stones composed of orotic acid. Following up on this observation, Demars et al. (1976) found that *spf*/Y mice have an abnormal form of liver ornithine carbamoyltransferase. The apparent net deficiency of the enzyme activity in affected males is about 90%. Normal and abnormal enzyme activity is found in the livers of heterozygous females. Qureshi et al. (1979, 1983), Briand et al. (1981), and Briand Cathelineau (1983) further characterized the enzyme defect. Hyperammonemia can be induced in *spf*/Y males (Spector and Mazzocchi 1982). The human disorder is highly variable, and Briand et al. (1982) were able to classify 5 groups of mutations based on kinetic and immunochemical studies of the enzyme. Two of these are similar to *spf* and *spf^{ash}* (see below) in enzymatic properties and the amounts of cross-reactive material. Ohtake et al. (1986) used a cloned rat OTC cDNA as a probe to isolate mRNA, its precursors in nuclei, and genomic DNA from *spf* and *spf^{ash}* mutants. The *spf* mutation is associated with a slightly decreased amount of OTC mature mRNA and with almost the same amount of nuclear mRNA precursors. Mullins and Chapman (1986) briefly mentioned the existence of at least 5 polymorphic forms of the gene. Veres et al. (1987) identified the point mutation underlying the sparse fur phenotype as C→A reversion at coding base 348, which results in the replacement of a histidine residue with an asparagine residue at amino acid position 117.

abnormal skin and hair, spf^ash   Hulbert and Doolittle (1971) described a mutant with abnormal skin and hair that proved to be an allele (*spf^ash*), of *spf* (Doolittle et al. 1974). Young homozygous females and hemizygous males have retarded hair growth and wrinkled skin. Adults are fully furred. The expression of the gene in proven heterozygotes varies considerably. In a footnote, Demars et al. (1976) say that a *spf^ash*/Y mouse possesses a second abnormal form of ornithine carbamoyltransferase. Although extracts of liver from *spf^ash*/Y males have only 5–10% of normal OTC activity, the homogeneous enzyme isolated from these extracts is identical to that of +/Y animals (Briand et al. 1983; Rosenberg et al. 1983). The mutation results in the formation of a reduced amount of translatable OTC mRNA that codes for the synthesis of 2 distinct OTC species; both of these are taken up and processed by mitochondria, but only the wild type is assembled to an active trimer. Rosenberg et al. speculate that "this novel phenotype may result from a mutation in the structural gene for OTC leading, primarily, to aberrant splicing of OTC messenger RNA, and secondarily, to formation of a structurally altered precursor whose post translational pathway is ultimately futile because its mature mitochondrial form is not capable of assembly and functional expression." Hodges et al. (1985) used cloned rat and human OTC cDNAs as probes, and dot blot and Northern blot hybridizations to analyze mRNA from the liver of mutant mice, and concluded that, if the allele affects mRNA splicing, either the mutation lies outside the sequences compared or is too subtle to be detected by the techniques used. Ohtake et al. (1986) found that the mutation is associated with a reduced amount of mature mRNA and of immunoreactive protein, but nuclear mRNA precursors do not accumulate. These authors believe this deficiency could be due to a mutation affecting either transcription of the gene, or stability of mRNA precursors, or both.

Rat

Takiguchi et al. (1987) described the rat OTC gene. It is 75 kb long and split into 10 exons. The introns range in length from 85 bases to 26 kb. The exons total 1.5 kb and occupy only 2% of the gene.

Briand P, Cathelineau L: 1983. Sparse-fur mutation: a model for some human ornithine transcarbamylase deficiencies. In: Lowenthal A, Mori A, Marescau B (eds), *Urea Cycle Diseases*. New York: Plenum Press, pp 185–194.

Briand P, Cathelineau L, Kamoun P, Gigot D, Penninckx M: 1981. Increase of ornithine transcarbamylase protein in sparse-fur mice with ornithine transcarbamylase deficiency. *FEBS Lett* 130: 62–68.

Briand P, François B, Rabier D, Cathelineau L: 1982. Ornithine transcarbamylase deficiencies in human males: kinetic and immunochemical classification. *Biochim Biophys Acta* 704: 100–106.

Briand P, Miura S, Mori M, Cathelineau L, Kamoun P, Tatibana M: 1983. Cell-free synthesis and transport of precursors of mutant ornithine carbamoyltransferases into mitochondria. *Biochem Biophys Acta* 760: 389–397.

Cupp WB: 1958. Private communication. *Mouse News Lett* 19: 37 only.

Demars R, LeVan SL, Trend BL, Russell LB: 1976. Abnormal ornithine carbamoyltransferase in mice having the sparse-fur mutation. *Proc Natl Acad Sci USA* 73: 1693–1697.

Doolittle DP, Hulbert LL, Cordy C: 1974. A new allele of the sparse fur gene in the mouse. *J Hered* 65: 194–195.

Grant SG, Stephenson DA, Chapman VM: 1987. Ornithine transcarbamylase expression in an extraembryonic cell lineage. *Mouse News Lett* 78: 74 only.

Hodges PE, Kraus JP, Rosenberg LE: 1985. Analysis of ornithine transcarbamylase mRNA in spf^{ash} mutant mice. (Abstr) *Am J Hum Genet* 37: A156 only.

Hulbert LL, Doolittle DP: 1971. Abnormal hair: a sex-linked mutation in the house mouse. *Genetics* 68: s29 only.

Kaslow DC, Migeon BR, Persico MG, Zollo M, VandeBerg JL, Samollow PB: 1987. Molecular studies of marsupial X chromosomes reveal limited sequence homology of mammalian X-linked genes. *Genomics* 1: 19–28.

Mullins LJ, Chapman V: 1986. Polymorphism of the ornithine transcarboxylase locus and control of gene expression. *Mouse News Lett* 74: 114 only.

Ohtake A, Takayanagi M, Yamamoto S, Kakinuma H, Nakajima H, Tatibana M, Mori M: 1986. Molecular basis of ornithine transcarbamylase deficiency in *spf* and *spf-ash* mutant mice. *J Inherited Metab Dis* 9: 289–291.

Qureshi IA, Letarte J, Ouellet R: 1979. Ornithine transcarbamylase deficiency in mutant mice. I. Studies on the characterization of enzyme defect and suitability as animal model of human disease. *Pediatr Res* 13: 807–811.

Qureshi IA, Letarte J, Ouellet R: 1983. Spontaneous animal models of ornithine transcarbamylase deficiency: studies on serum and urinary nitrogenous metabolites. In: Lowenthal A, Mori A, Marescau B (eds), *Urea Cycle Disease*. New York: Plenum Press, pp 173–183.

Rosenberg LE, Kalousek F, Orsulak MD: 1983. Biogenesis of ornithine transcarbamylase in spf^{ash} mutant mice: two cytoplasmic precursors, one mitochondrial enzyme. *Science* 222: 426–428.

Russell LB: 1960. Private communication. *Mouse News Lett* 22: 50 only.

Scherer SE, Veres G, Caskey CT: 1988. The genetic structure of mouse ornithine transcarbamylase. *Nucleic Acids Res* 16: 1593–1601.

Sinclair AH, Wrigley JM, Graves JAM: 1987. Autosomal assignment of OTC in marsupials and monotremes: implications for the evolution of sex chromosomes. *Genet Res* 50: 131–136.

Spector EB, Mazzocchi RA: 1982. An animal model for a human inborn error of metabolism. *Am J Hum Genet* 34: 63A only.

Takiguchi M, Murakami T, Miura S, Mori M: 1987. Structure of the rat ornithine carbamoyltransferase gene, a large X chromosome-linked gene with an atypical promoter. *Proc Natl Acad Sci USA* 84: 6136–6140.

Veres G, Gibbs RA, Scherer SE, Caskey CT: 1987. The molecular basis of the sparse fur mutation. *Science* 237: 415–417.

*31135 OUABAIN RESISTANCE, X-LINKED (OUBR)

Mouse, *Oubr*

Ouabain, a steroidal glycoside with digitalis-like action, inhibits plasma membrane Na/K-ATPase. Compared to human cells, mouse cells are relatively resistant to ouabain. Law et al. (1984) used somatic cell hybrids to indicate that a gene(s) controlling this resistance is syntenic with and may be closely linked to *Hprt* (30800).

Law ML, Mo X, Zhang X, Kao FT: 1984. Genes coding for ouabain resistance (OUBR) and HPRT are syntenic in the mouse genome. (Abstr) *Cytogenet Cell Genet* 37: 518 only.

*31147 PARALYTIC TREMOR [?31150]

Rabbit, *pt*

This disorder, first noted among a colony of rabbits in the Department of Laboratory Animal Breeding of the Polish Academy of Science in Lomma-Las, was originally described by Osetowska and Wiśniewski (1966) and Osetowska (1967). Affected males and affected females (produced by crossing carrier females with affected males) are recognized during the first week after birth by the presence of a coarse tremor resembling that of parkinsonism in man; a progressive spastic paralysis of the limbs develops by 4–6 weeks. The clinical course varies considerably. In about 40% of cases, the disorder is severe and paralysis complete; incontinence of bowel and bladder develops and the tremors increase. These animals die at 3–7 months. In other animals, the disease is milder and involves increased muscle tone and exaggerated tendon reflexes. The disease can become stationary and even improve somewhat; animals in which this happens can reproduce. Neuropathological changes involve neuroaxonal degeneration followed by demyelination and glial fibrosis. The lesions involve all parts of the nervous system, but are most apparent in the basal ganglia, cord, cerebellum, medulla, and cerebral cortex. There are 2 types of neuronal lesions. The first includes swelling, chromatolysis, and removal of large and small neurons; there is often an associated edema of the oligodendroglia. The second lesion, found in close to 60% of brains studied histologically, was termed pseudocalcification. The 2 lesions together are considered specific for the disease (Osetowska 1967). The symbol *pt* was assigned by Lindsey and Fox (1974) after consultation with Osetowska. Beginning in the mid 1970s (Osetowska and Luszawski 1975), Osetowska and her colleagues published a series of papers further delineating the disorder. In the third report, Osetowska et al. (1975) described a high frequency (40%) of degeneration of retinal neurons, and impairment of myelin and nerve fibers in the optic nerves, chiasma, and tracts. In the sixth report, Osetowska et al. (1976) described the neuronal calcification in more detail. Subsequent reports have appeared sporadically; the latest of which I am aware was by Taraszewska (1986). Despite much effort, the fundamental pathogenic error in this disorder remains unknown and its relation to the X-linked form of parkinsonism unclear.

Lindsey JR, Fox RR: 1974. Inherited diseases and variations. In: Weisbroth SH, Flatt RE, Krauss AL (eds), *The Biology of the Laboratory Rabbit.* New York: Academic Press, pp 380–381.

Osetowska E: 1967. Nouvelle maladie héréditaire du lapin de laboratoire. *Acta Neuropathol (Berl)* 8: 331–344.

Osetowska E, Luszawski F: 1975. I. "Prolegomena" to experimental studies on a hereditary disease of nervous system on the "pt" rabbit model. (Pol, Eng summary) *Neuropatol Pol* 13: 61–70.

Osetowska E, Wiśniewski H: 1966. Ataxie familiale du lapin, différente de la maladie heféditaire de Sawin–Anders: première communication. *Acta Neuropathol (Berl)* 6: 243–250.

Osetowska E, Taraszewska A, Luszawski F: 1975. III. Changes in the optic system of "pt" rabbit (preliminary report). (Pol, Eng summary) *Neuropatol Pol* 13: 87–92.

Osetowska E, Taraszewska A, Luszawski F: 1976. VI. Neuronal calcification in "pt" rabbits: dependence on age and clinical course. (Pol, Eng summary) *Neuropatol Pol* 14: 85–95.

Taraszewska A: 1986. Morphological characteristic of sudanophilic deposits appearing in the course of abnormal myelination in pt rabbit. (Pol, Eng summary) *Neuropathol Pol* 24: 183–194.

Taraszewska A, Zelman IB: 1981. Characteristics of the topography of changes in *pt* rabbit brain during development of the chinical symptoms in the chronic course of the disease. (Pol, Eng summary) *Neuropatol Pol* 19: 441459.

*31156 PATCHY COAT [NK]

Mouse

Bode et al. (1987) briefly described 4 independent mutations, induced with ethylnitrosourea, which produce a similar phenotype in heterozygous females. The mutations are lethal in males, and heterozygous females have transverse striping due to absence of hairs in the dark stripes. Expressivity in females is variable; some appear normal but others die by 3 weeks of age.

Bode VC, Justice MJ, McDonald JD: 1987. Patchy coat. *Mouse News Lett* 77: 133 only.

*31180 PHOSPHOGLYCERATE KINASE 1 (PGK1; EC 2.7.2.3)

This is the structural locus of the isozyme of PGK found in somatic tissues.

African green monkey

Garver et al. (1978) used somatic cell hybridization to demonstrate X-linkage.

American mink

Rubtsov et al. (1982) used somatic cell hybridization to demonstrate X-linkage.

Brush-tailed possum (*Trichosurus vulpecula*)

VandeBerg et al. (1979) described X-linkage of 2 rare PGK–1 variants in this Australian marsupial. Dosage compensation occurs through paternal X-inactivation.

Cattle

X-linkage was demonstrated by Heuertz and Hors-Cayla (1978) using somatic cell hybridization, and was confirmed by Shimizu et al. (1981).

Chimpanzee

Chen et al. (1976) used somatic cell hybridization to demonstrate synteny with *G6PD*, and Cochet et al. (1977) and Garver et al. (1978) independently demonstrated X-linkage.

Chinese hamster

Westerveld et al. (1972) used gene dosage studies to show that the locus was linked to those for HPRT and G6PD in Chinese hamster cells.

Dasyurids

In these marsupials, the locus is syntenic with *HPRT*, *G6PD*, and *GLA* (Dobrovic and Graves 1986); in 2 species, *Sminthopsis crassicaudata* and *Planigales maculata*, the gene order appears to be *G6PD–HPRT–PGK–GLA*, with the first-named locus well separated from the others.

Gibbon

Turleau et al. (1983) used somatic cell hybridization to demonstrate X-linkage.

Gorilla

Garver et al. (1978) used somatic cell hybridization to demonstrate X-linkage.

Horse

Ohno (1973) refers to an unpublished thesis by B. F. Deys in which X-linkage is demonstrated. Banks et al. (1979) determined the amino acid sequence and the structure of the enzyme from horse muscle.

Indian muntjac

Shows et al. (1976) used somatic cell hybridization to demonstrate X-linkage. The locus is on Xp (Ropers et al. 1982)

Macropodids (kangaroos and wallabies)

Cooper et al. (1971) found PGK polymorphism in several species and subspecies of kangaroos, and demonstrated X-linkage in 3 species of *Macropus* (the great kangaroos). Dosage compensation occurs through paternal X-inactivation. Graves et al. (1979) used marsupial–mouse cell hybrids to demonstrate that the genes for G6PD, HPRT, and PGK are X-linked, and Donald and Hope (1981) used hybrid cell studies to localize the 3 genes to the terminal portion of the euchromatic arm of the red kangaroo X chromosome. Dawson and Graves (1984) showed that the gene order is *G6PD–HPRT–PGK* in the gray kangaroo, and the same authors (Dawson and Graves 1986) demonstrated the same gene order on the long arm of the euro and wallaroo X chromosomes. On the long arm of the wallaroo X, *GLA* is located between *HPRT* and *G6PD*. It appears that all *Macropus* species share a common segment bearing these loci.

Mouse, *Pgk-1*

Kozak et al. (1974) demonstrated X-linkage of the locus in the mouse by comparing PGK activity levels in oocytes from XX and XO females. *Mus musculus* and *M. caroli*, a related species from Thailand, have different electrophoretic mobilities for a number of enzymes, including PGK; Chapman and Shows (1976) used fetuses obtained by interspecific hybridization through artificial insemination to confirm X-linkage. Nielsen and Chapman (1977) provided direct evidence of X-linkage, using electrophoretic variants found as a polymorphism in feral mice in Denmark; the allele in these mice is designated *Pgk-1*^{a}, and that in inbred mice and other wild mice, *Pgk-1*^{b}. Heterozygotes have both bands. Muhlbacher et al. (1983) purified the 2 allelic isozymes and found them to be identical in all properties examined except specific activity and thermostability; the 1B allozyme has higher values of both. The locus is in the midregion of the chromosome, clustered with several other loci. In early mouse embryos, the maternally derived allele is already activated during late preimplantation development, whereas the paternally derived allele is silent at this stage but is subsequently expressed at approximately the time of X-chromosal activation (Kriestsch et al. 1982). Mori et al. (1986) determined the nucleotide sequence of a cDNA clone containing the entire coding region of the gene. There is 93% similarity with the human gene. The resultant enzyme contains 416 amino acids and is 98% and 96% similar to the human and horse enzymes, respectively. There is about 60% similarity with the yeast and gene and enzyme. Adra et al. (1987) cloned the gene and determined the nucleotide sequence of its promoter; the sequence is very similar to that of the human *PGK* promoter.

Mouse lemur

Cochet et al. (1982) used somatic cell hybridization to demonstrate X-linkage.

Owl monkey, *Aotus*

Ma (1983) used somatic cell hybridization to demonstrate X-linkage.

Pig

Gellin et al. (1980) used somatic cell hybridization to demonstrate synteny with *G6pd* and *Hprt*.

Rabbit

Echard and Gillois (1979) and Cianfrigilia et al. (1979) used somatic cell hybridization to demonstrate that the locus is syntenic with *Hprt*, *G6pd*, and *Gla*; Echard et al. (1981) localized it on the X chromosome.

Rat

Yoshida (1978) used somatic cell hybridization to demonstrate X-linkage.

Sheep

Saïdi-Mehtar et al. (1981) used somatic cell hybridization to demonstrate synteny with *G6pd* and *Gala*.

Virginia opossum (*Didelphis virginiana*), *Pgk-A*

Samollow et al. (1987) demonstrated that the locus is X-linked in this American marsupial. The paternal allele is preferentially inactivated, as it is in Australian marsupials. Although dosage compensation appears to be complete for this locus, it is not for all; see *G6pd* (30590).

Adra CN, Boer PH, McBurney MW: 1987. Cloning and expression of the mouse *pgk-1* gene and the nucleotide sequence of its promoter. *Gene* 60:65–74.

Banks RD, Blake CCF, Evans PR, Haser R, Rice DW, Hardy GW, Merrett M, Phillips AW: 1979. Sequence, structure and activity of phosphoglycerate kinase: a possible hinge-bending enzyme. *Nature* 279: 773–777.

Chapman VM, Shows TB: 1976. Somatic cell genetic evidence for X-chromosome linkage of three enzymes in the mouse. *Nature* 259: 665–667.

Chen S, McDougall JK, Creagan RP, Lewis V, Ruddle FH: 1976. Mapping of genes and adenovirus-12-induced gaps using chimpanzee–mouse somatic cell hybrids. *Cytogenet Cell Genet* 16:412–415.

Cianfriglia M, Miggiano VC, Meo T, Muller HJ, Muller E, Battistuzzi G: 1979. Evidence for synteny between the rabbit *HPRT*, *PGK* and *G6PD* in mouse × rabbit somatic cell hybrids. (Abstr) *Cytogenet Cell Genet* 25: 142 only.

Cochet C, Finaz, Van Cong N, Frezal J, de Grouchy J: 1977. Localisation des gènes de la beta-glycuronidase, de l'énolase-2 et de la phosphoglycerate kinase sur les chromosomes 7, 12 et X du chimpanze. *Ann Génét* 20: 255–257.

Cochet C, Creau-Goldberg N, Turleau C, de Grouchy J: 1982. Gene mapping of *Microcebus murinus* (Lemuridae): a comparison with man and *Cebus capucinus* (Cebidae). *Cytogenet Cell Genet* 33: 213–221.

Cooper DW, VandeBerg JL, Sharman GB, Poole WE: 1971. Phosphoglycerate kinase polymorphism in kangaroos provides further evidence for paternal X inactivation. *Nature [New Biol]* 230: 155–157.

Cooper DW, Wooley PA, Maynes GM, Sherman FS, Poole WE: 1983. Studies on metatherian sex chromosomes. XII. Sex-linked inheritance and probable paternal X-inactivation of alpha-galactosidase A in Australian marsupials. *Aust J Biol Sci* 36: 511–517.

Dawson GW, Graves JAM: 1984. Gene mapping in marsupials and monotremes. I. The chromosomes of rodent–marsupial (*Macropus*) cell hybrids, and gene assignments to the X chromosome of the grey kangaroo. *Chromosoma* 91: 20–27.

Dawson GW, Graves JAM: 1986. Gene mapping in marsupials and monotremes. III. Assignment of four genes to the X chromosome of the wallaroo and the euro (*Macropus robustus*). *Cytogenet Cell Genet* 42: 80–84.

Dobrovic A, Graves JAM: 1986. Gene mapping in marsupials and monotremes. II. Assignments to the X chromosome of dasyurid marsupials. *Cytogenet Cell Genet* 41: 9–13.

Donald JA, Hope RM: 1981. Mapping a marsupial X chromsome using kangaroo–mouse somatic cell hybrids. *Cytogenet Cell Genet* 29: 127–137.

Echard G, Gillois M: 1979. *G6PD–PGK–GLA–HPRT* synteny in the rabbit, *Oryctolagus cuniculus*. (Abstr) *Cytogenet Cell Genet* 25: 148–149.

Echard G, Gellin J, Benne F, Gillois M: 1981. The gene map of the rabbit (*Oryctolagus cuniculus L.*). I. Synteny between the rabbit gene loci coding for HPRT, PGK, G6PD, and GLA: their localization on the X chromosome. *Cytogenet Cell Genet* 29: 176–183.

Garver JJ, Pearson PL, Estop A, Dijksman TM, Wijnen LMM, Westerveld A, Meera Khan P: 1978. Gene assignments to the presumptive homologs of human chromosomes 1, 6, 11, 12, and X in the Pongidae and Cercopithecoidea. *Cytogenet Cell Genet* 22: 564–569.

Gellin J, Benne F, Hors-Cayla MC, Gillois M: 1980. Carte génique du porc (*Sus scrofa l.*). I. Etude de deux groupes synténiques G6PD, PGK, HPRT et PKM2, MPI. *Ann Genet* 23: 15–21.

Graves JAM, Chew GK, Cooper DW, Johnston PG: 1979. Marsupial–mouse cell hybrids containing fragments of the marsupial X chromosomes. *Somat Cell Genet* 5:481–489.

Heuertz S, Hors-Cayla M-C: 1978. Carte génétique des bovins par la technique d'hybridation cellulaire. Localisation sur le chromosome X de la glucose-6-phosphate déshydrogénase, la phosphoglycérate kinase, l'alpha-galactosidase A et l'hypoxanthine guanine phos-phoribosyl transferase. *Ann Génét* 21: 197–202.

Kozak LP, McLean GK, Eicher EM: 1974. X linkage of phosphoglycerate kinase in the mouse. *Biochem Genet* 11: 41–47.

Krietsch WKG, Fundele R, Kuntz GWK, Fehlau M, Burki K, Illmensee K: 1982. The expression of the X-linked phosphoglycerate kinase in the early mouse embryo. *Differentiation* 23: 141–144.

Ma NSF: 1983. Gene map of the New World Bolivian owl monkey, *Aotus*. *J Hered* 74: 27–33.

Mori N, Singer–Sam J, Lee C-Y, Riggs AD: 1986. The nucleotide sequence of a cDNA clone containing the entire coding region for mouse X-chromosome-linked phosphoglycerate kinase. *Gene* 45: 275–280.

Mühlbacher C, Kuntz GWK, Haedenkamp GA, Krietsch WKG: 1983. Comparison of the two purified allozymes (1B and 1A) of X-linked phosphoglycerate kinase in the mouse. *Biochem Genet* 21: 487–496.

Nielsen JT, Chapman VM: 1977. Electrophoretic variation for *X*-chromosome-linked phosphogycerate kinase (PGK-1) in the mouse. *Genetics* 87: 319–325.

Ohno S: 1973. Ancient linkage groups and frozen accidents. *Nature* 244: 259–262.

Ropers HH, Sperling K, Raman R, Schmelzer B, Stromaier U, Neitzel H: 1982. Indian muntjac: gene assignments on the short and the long arm of the X chromosome. (Abstr) *Cytogenet Cell Genet* 32: 312 only.

Rubtsov NB,Radjabli SI, Gradov AA, Serov OL: 1982. Chromosome localization of the genes for isocitrate dehydrogenase-1, isocitrate dehydrogenase-2, glutathione reductase, and phosphoglycerate kinase-1 in the American mink (*Mustela vison*). *Cytogenet Cell Genet*33: 256–260.

Saïdi-Mehtar N, Hors-Cayla M-C, Van Cong N: 1981. Sheep gene mapping by somatic cell hybridization: four syntenic groups: ENO1–PGD, ME1–PGM 3, LDHB–PEPB–TPI, and G6PD–PGK–GALA. *Cytogenet Cell Genet* 30: 193–204.

Samollow PB, Ford AL, VandeBerg JL: 1987. *X*-linked gene expression in the Virginia opossum: differences between the paternally derived *Gpd* and *Pgk-A* loci. *Genetics* 115: 185–195.

Shimizu N, Shimizu Y, Kondo I, Woods C, Wegner T: 1981. The bovine genes for phosphoglycerate kinase, glucose-6-phosphage dehydrogenase, alpha-galactosidase, and hypoxanthine phosphoribosyltransferase are linked to the X-chromosome in cattle–mouse cell hybrids. *Cytogenet Cell Genet* 29: 26–31.

Shows TB, Brown JA, Chapman VM: 1976. Comparative gene mapping of *HPRT, G6PD*, and *PGK* in man, mouse, and muntjac deer. *Cytogenet Cell Genet* 16: 436–439.

Turleau C, Creau-Goldberg N, Cochet C, de Grouchy J: 1983. Gene mapping of the gibbon. Its position in primate evolution. *Hum Genet* 64: 65–72.

VandeBerg JL, Thiel JE, Hope RM, Cooper DW: 1979. Expression of PGK-A in the Australian brush-tailed possum, *Trichosurus vulpecula* (Kerr), consistent with paternal X-inactivation. *Biochem Genet* 17: 325–332.

Westerveld A, Visser RPLS, Freeke MA, Bootsma D: 1972. Evidence for linkage of 3-phosphoglycerate kinase, hypoxanthine–guanine–phosphoribosyltransferase, and glucose 6-phosphate dehydrogenase loci in Chinese hamster cells studied by using a relationship between gene multiplicity and enzyme activity. *Biochem Genet* 7: 33–40.

Yoshida MC: 1978. Rat gene mapping by rat–mouse somatic cell hybridization and comparative Q-banding analysis between rat and mouse chromosomes. *Cytogenet Cell Genet* 22: 606–609.

*31187 PHOSPHORYLASE KINASE (DEPHOSPHORYLASE KINASE; ATP: PHOSPHORYLASE-b PHOSPHOTRANSFERASE; PHK; EC 2.7.1.38) [30600]

This locus controls the activity of phosphorylase kinase, which catalyzes the conversion of phosphorylase *b* to phosphorylase *a*, which initiates glycogenolysis in muscle. The enzyme in skeletal muscle is composed of 4 subunits – alpha, beta, gamma, and delta (Cohen et al. 1978).

Mouse, *Phk*

Lyon and Porter (1962, 1963) described a strain of mice, I/FnLn, in which phosphorylase in skeletal muscle is always in the *b* form presumably because the enzyme is absent. The I strain possesses an allele, *Phk^b*, which determines almost complete absence of the enzyme in skeletal muscle and reduced amounts in brain, heart, and kidney. In the muscle, the enzyme deficiency is accompanied by a 3- to 5-fold elevation in glycogen and a decreased rate of glycogen breakdown. All other strains possess the *Phk^a* allele which determines normal levels of the enzyme. In heterozygous females, the enzyme activity varies considerably (Lyon 1970). Lyon et al. (1967) used the I strain in crosses to show that *Phk* is X-linked. The locus is in the mid region of the chromosome (Huijing et al. 1973). In man, a deficiency of phosphorylase kinase produces an extremely mild form of glycogen storage disease. Huijing (1970a) compared the human and murine defects; in neither species does the defect result in significant disease. Although the enzyme defect is strikingly similar in the 2 species, there are some differences (Huijing 1970b). Cohen and Cohen (1973) found that, although I-strain mice have essentially no phosphorylase activity, they contain a mutant protein which is present in concentrations similar to that of the normal enzyme and is immunologically indistinguishable from it. The conclusion by Gross et al. (1975) that the *Phk^b* allele results in normal amounts of an enzyme with abnormal structure was disputed by Cohen et al. (1976), who were unable to demonstrate that I-strain muscle possessed any of the 4 polypeptide chains comprising PHK. The latter authors concluded that the *Phk* X-linked locus controls the expression of structural genes for the 4 chains; at least 1 of these genes is located on an autosome. Lyon (1970) detected small amounts of enzyme activity during the first 3 weeks of life; Daegelen-Proux et al. (1978) demonstrated that this is caused by "fetal form" of the enzyme, the activity of which decreases after birth. Varsanyi et al. (1978) found nearly normal levels of activity in sarcoplasmic reticulum isolated from I-strain mice. Varsanyi et al. (1980) described a partial deficiency of phosphorylase kinase caused by an allele, *Phk^c*, whose effect is dominant over those of the other alleles. Chamberlain et al. (1987) isolated and characterized cDNA clones for the gamma subunit of the enzyme (gamma-Phk). Muscle from animals of the Phk-deficient I strain contains reduced levels of gamma-Phk mRNA. However, hybridization experiments indicate that the gamma-*Phk* gene is not on the X chromosome. The reduced levels of the gamma subunit appear to be an effect of the deficient phenotype and do not stem from a mutation of the subunit gene. The total amount of calmodulin, the delta subunit in the phosphorylase kinase hexadecamer, is reduced by 40% in tissue deficient in phosphorylase kinase (Picton et al. 1983). This reduction results from a reduction in the abundance of calmodulin mRNA (Bender et al. 1988).

Bender PK, Dedman JR, Emerson CP Jr: 1988. The abundance of calmodulin mRNAs is regulated in phosphorylase kinase-deficient skeletal muscle. *J Biol Chem* 263: 9733–9737.

Chamberlain JS, VanTuinen P, Reeves AA, Philip BA, Caskey CT: 1987. Isolation of cDNA clones for the catalytic gamma subunit of mouse muscle phosphorylase kinase: expression of mRNA in normal and mutant *Phk* mice. *Proc Natl Acad Sci USA* 84:2886–2890.

Cohen PTW, Cohen P: 1973. Skeletal muscle phosphorylase kinase deficiency: detection of a protein lacking any activity in ICR/IAn mice. *FEBS Lett* 29: 113–116.

Cohen PTW, Burchell A, Cohen P: 1976. The molecular basis of skeletal muscle phosphorylase kinase deficiency. *Eur J Biochem* 66: 347–356.

Cohen P, Burchell A, Foulkes JG, Cohen PTW, Vanaman TC, Nairn AC: 1978. Identification of the Ca^{2+}-dependent modulator protein as the fourth subunit of rabbit skeletal muscle phosphorylase kinase. *FEBS Lett* 92: 287–293.

Daegelen–Proux D, Alexandre Y, Dreyfus J-C: 1978. Phosphorylase kinase isoenzymes in deficient ICR/IAn mice. *Eur J Biochem* 90: 369–375.

Gross SR, Longshore MA, Pangburn S: 1975. The phosphorylase kinase deficiency (*Phk*) locus in the mouse: evidence that the mutant allele codes for an enzyme with an abnormal structure. *Biochem Genet* 13: 567–584.

Huijing F: 1970a. Phosphorylase kinase deficiency. *Biochem Genet* 4: 187–194.

Huijing F: 1970b. Phosphorylase kinase deficiency in mice. *FEBS Lett* 10: 328–332.

Huijing F, Eicher EM, Coleman DL: 1973. Location of phosphorylase kinase (*Phk*) in the mouse X chromosome. *Biochem Genet* 9: 193–196.

Lyon JB Jr: 1970. The X-chromosome and the enzymes controlling muscle glycogen phosphorylase kinase. *Biochem Genet* 4: 169–185.

Lyon JB Jr, Porter J: 1962. The effect of pyridoxine deficiency on muscle and liver phosphorylase of two inbred strains of mice. *Biochem Biophys Acta* 58: 248–254.

Lyon JB Jr, Porter J: 1963. The relation of phosphorylase to glycogenolysis in skeletal muscle and heart of mice. *J Biol Chem* 238: 1–11.

Lyon JB Jr, Porter J, Robertson M: 1967. Phosphorylase *b* kinase inheritance in mice. *Science*155: 1550–1551.

Picton C, Shenolikar S, Grand R, Cohen P: 1983. Calmodulin as an integral subunit of phosphorylase kinase from rabbit skeletal muscle. *Methods Enzymol* 102: 219–227.

Varsanyi M, Groschel-Stewart U, Heilmeyer LMG Jr: 1978. Characterization a Ca^{2+}-dependent protein kinase in skeletal muscle membranes of I-strain and wild-type mice. *Euro J Biochem* 87: 331–340.

Varsanyi M, Vrbica A, Heilmeyer LMG Jr: 1980. X-linked dominant inheritance of partial phosphorylase kinase deficiency in mice. *Biochem Genet* 18: 247–261.

31201 POLYDACTYLY

Cattle

Morrill (1945) reported the occurrence of polydactyly affecting only the forelimbs in Hereford cattle in Utah. The meager breeding data – 2 presumptive carrier females and 4 offspring – suggest X-linkage. To my knowledge no further reports on this trait have been published.

Morrill EL: 1945. A new sex-linked defect in cattle. *J Hered* 36: 81–821.

*31202 POLYDACTYLY, PREAXIAL, WITH HEMIMELIA AND UROGENITAL DEFECTS, X-LINKED [NK]

Mouse, *Xpl*

This mutation, which arose spontaneously at the Jackson Laboratory, was described by Sweet and Lane (1980). The trait, recognizable in most affected animals at birth, is a dominant with 93% penetrance in hemizygotes and 96% in heterozygotes. Expressivity varies from a barely discernible thickened hallux or elongated toe on 1 foot to up to 4 extra toes on 1 or both feet; tibial hemimelia, which may cause twisting on 1 or both feet; tibial hemimelia, which may cause twisting of 1 or both feet, occurs in approximately 42% of hemizygotes and 12% of heterozygotes. Urogenital anomalies including hydroureter, hydronephrosis, and cystic or missing kidneys are present in some hemizygotes; no similar defects have been reported in heterozygotes. Most affected females are fertile, but affected males are commonly sterile, probably because the testes fail to descend. Heterozygotes and homozygotes are phenotypically indistinguishable. Linkage studies indicated that the locus is located distally in a cluster with *Hyp*, *Gy*, and *Li*. No human homology has been discribed. Chase (1946) described a preaxial polydactyly that occurred as an isolated defect affecting the right hind food predominantly. Although he originally claimed the trait was a sex-linked recessive, this could not be substantiated (Chase, personal communication quoted by Grüneberg 1952).

Chase HB: 1946. A sex-linked recessive in the mouse. (Abstr) *Genetics* 31: 214 only.
Grüneberg H: 1952. *The Genetics of the Mouse*, 2nd ed. The Hague: Martinus Nijhoff, p 311.
Sweet HO, Lane PW: 1980. X-linked polydactyly (*Xpl*), a new mutation in the mouse. *J Hered* 71: 207–209.

*31208 PROTEOLIPID PROTEIN, MYELIN (LIPOPHILIN; PLP) [?31160]

The myelin of the CNS contains 2 major membrane proteins; myelin basic protein and proteolipid protein (lipophilin). The amino acid sequence of the human and bovine lipophilin is completely conserved (Stoffel et al. 1985), and the rat protein differs from them in only 3 among 276 amino acid residues (Jolles et al. 1983; Kronquist et al. 1987). The mRNA coding for the complete rat lipophilin was isolated and sequenced by Dautigny et al. (1985) and Milner et al. (1985). Nave et al. (1987a) showed that rat and mouse brain PLP gene transcripts exist in 2 alternatively spliced forms. When the complete third exon of the PLP gene is included after splicing, a PLP mRNA results, which encodes the 276-amino acid major PLP. The alternatively spliced mRNA, expressed with about 50% abundance relative to PLP mRNA lacks the 3' part of the third exon and encodes the sequences of a DM-20, a related 241-amino acid proteolipid. Campagnoni and Macklin (1988) reviewed the molecular biology of all the meylin proteins, including PLP. The human and mouse genes, in which the exon–intron arrangements are similar (Diehl et al. 1986; Macklin et al. 1987; Moriguchi et al. 1987; Ikenaka et al. 1988), have been localized to the X chromosome (Willard and Rirordan 1985); and the evidence for X-linkage in the rat is convincing (Naismith et al. 1987). The 3 mutations described below – shaking-pup disease in the dog, jimpy in the mouse, and myelin deficiency in the rat – phenotypically resemble Pelizaeus-Merzbacher disease (PMD) (MIM 31160) in the human. The findings that PLP is absent or severely reduced in the central nervous tissue of the nonhuman mutants, that *Plp* is X-linked in the human and mouse, that a mutation at the *Plp* locus underlies the disease in the rat, and that PLP was absent from the brain of an 18-year-old patient with PMD provide compelling evidence that all 4 diseases result from homologous mutations at the PLP locus. For details, see below.

Dog, shaking pups

Griffith et al. (1981a,b) described this disorder in male Springer spaniel pups in 3 litters from the same bitch; the first litter was sired by 1 dog, and the other 2 by another. Among 16 pups that lived to be studied adequately, there were 7 normal females, 3 normal males, and 6 affected males. Subsequently, 1 of the normal females produced affected males (Duncan et al. 1983). Affected animals are extremely small, show gross generalized tremors (particularly when aroused) at about 10–12 days of age, and die within 3–4 months. The CNS is severely hypomyelinated throughout, and axons are either naked or surrounded by an abnormally thin layer of myelin; the PNS is myelinated normally. The number of oligodendrocytes is severely reduced and many of them are abnormal (Duncan et al. 1983). PLP and the related DM-20 protein are expressed in approximately equal and greatly reduced amounts in the spinal cord of affected pups (Yanagisawa et al. 1987). This observation, together with the striking similarity of the disorder to the jimpy mouse and myelin-deficient rat (see below), makes it reasonable to assume that the shaking-pup disorder results from a mutation in the *Plp* gene. Female heterozygotes show myelin mosaicism of the optic nerve and spinal cord. Abnormal oligodendrocytes with distended rough endoplasmic reticulum are found in the abnormal myelin patches. Some of these females develop a marked tremor that disappears with age (Duncan et al. 1987a).

Mouse, *Plp*

Willard and Riordan (1985) used cloned bovine lipophilin cDNA to show that the locus is X-linked in both man and mouse. It encodes 2 different but related mRNA transcripts, PLP and DM-20, an essentially identical protein. In man, the locus maps to Xq22, a region that contains *PGK* and *GLA* (Mattei et al. 1986). In the mouse, the locus maps "very close" to *Ags* (*GLA*) (30105). Willard and Riordan (1985) suggested that mutations at the lipophilin locus in mouse and man may be responsible for jimpy and the Pelizaeus–Merzbacher disease, respectively. This suggestion has been proved correct for *jp* (see below), and Koeppen et al. (1987) reported evidence to support it in the human.

jimpy, jp   This mutant was first described by Phillips (1954); 32 years later, Daugtigny et al. (1986) and Nave et al. (1986) demonstrated that *jp* is a mutation at the structural locus coding for myelin proteolipid protein. For 10 years after it was described, the mutant received little attention, but, following descriptions of the first pathological details by Sidman et al. (1964, 1965), a prodigious number of reports on various aspects appeared; only a few of these will be mentioned here. Hogan and Greenfield (1984) provided an excellent review. Affected males have a generalized tremor, recognizable at 10–21 days of age, which is most marked in the hindquarters; by weaning age, a few animals have complete paralysis of the hind limbs. By the third or fourth week, the animals develop tonic–clonic seizures without focal onset and usually die by 30 days, commonly after a seizure. There is a severe deficiency of myelin formation in the CNS; the PNS is unaffected. The white matter of the CNS is almost completely absent, and hardly any axons show more than a few sheaths of uncompacted myelin. Many biochemical abnormalities have been investigated in great detail; the most significant involve a drastic decrease of structural myelin proteins and enzymes concerned with myelin lipid metabolism (summarized by Hogan and Greenfield 1984). The results of these and other morphological studies (Wolf and Holden 1969; Privat et al. 1972; Meier and Bischoff 1975; Kanpp et al. 1986) indicated that the primary pathogenic defect is expressed in the oligodendrocytes. The gene is expressed biochemically and morphologically in heterozygous females (Skoff and Nowicki Montgomery 1981; Hatfield and Skoff 1982; Benjamins et al. 1984, 1986; Kerner and Carson 1984; Bartlett and Skoff 1986). The myelin content is reduced in young heterozygotes (Rosenfeld and Friedrich 1984) but recovers to normal in older heterozygotes as a consequence of an increased rate of oligodendrocyte production (Rosenfeld and Friedrich 1986). Tsuji (1985) successfully treated affected animals by administering forphenicinol, a low molecular weight immunomodifier, during the first few days after birth. Continuous treatment with 1.0 mg/day beginning 2-6 days after birth resulted in a striking recovery of myelin-associated 2′,3′-cyclic nucleotide 3′-phosphodydrolase and cholesterol-ester hydrolase activity and of total myelin concentration of the CNS; a complete remission of the intention tremor and the convulsive seizures in 8 of 13 treated animals, whose life-span was markedly prolonged; and recovery to normal of galactocerebroside synthesis in the CNS. Attention came to focus on lipophilin as the key protein involved in the disorder because it is expressed specifically in the CNS (Lees and Brostoff 1984) and is the most severely reduced myelin component in the brains of affected animals (Nussbaum and Mandel 1973; Lerner et al. 1974; Matthieu et al. 1974; Kerner and Carson 1984; Sorg et al. 1986; Gardinier et al. 1986; Yanagisawa and Quarles 1986). This focus sharpened when Willard and Riorden (1985) found that PLP was X-linked in mouse and man. Nave et al. (1986) showed that the amount of PLP mRNA is markedly reduced in the brains of jimpy mice, and that cDNA clones encoding jimpy PLP contain a 74-base deletion within the coding region; Morello et al. (1986) independently made the same observation. However, the sequence that is absent from the jimpy PLP mRNA is fully preserved in the jimpy *PLP* gene. These authors and Hudson et al. (1987) suggested independently that *jp* involves a primary lesion in the *PLP* gene that results in an incorrectly spliced RNA transcript. Subsequently Nave et al. (1987b) showed that the missing segment corresponds to a separate exon, equivalent to exon 5 in the human *PLP*, and that the structure of a 3′ splice acceptor site upstream of that exon deviates from a 100% conserved consensus splice site. They conclude that the primary genetic defect in jimpy is a single base change, A→G, in the 3′ acceptor splice site that disables an invariant recognition of RNA splicing and leads to the deletion of exon 5. Similar findings were reported by Ikenaka et al. (1988). Moriguchi et al. (1987) used an RNase mapping procedure to show also that the deletion corresponds exactly to the fifth exon. The proteolipids appear to be synthesized in the rough endoplasmic reticulum and transported via the Golgi apparatus toward the forming myelin sheath. Roussel et al. (1987) used immunocytochemical techniques to demonstrate that, although the synthesis of the proteolipid molecules takes place normally in the rough endoplasmic reticulum of the CNS of jimpy mice, further processing that requires them to be incorporated in the Golgi membranes is inhibited; presumably this inhibition is related to the abnormal structure of the mutant molecules. Because of the similarity of the pathological changes in jimpy and the Pelizaeus–Merzbacher disease (PMD) (MIM 31160), Torii et al. (1971) suggested homology between the two. Lyon and Goffinet (1980) state that jimpy has many features in common with the Seitelberger subtype (Seitelberger 1954). The findings of Willard and Riordan (1985), noted above, and those of Koeppen et al. (1987) that PLP was absent from the brain of an 18-year-old patient with PMD provide strong evidence to support the suggestion that PMD and jimpy result from homologous mutations at the locus determining PLP.

myeline synthesis deficiency, jp^{msd}   This mutant was first described in detail by Meier and MacPike (1970). Affected males had tremors and repeated seizures, and died at 18–23 days of age; the earliest signs appeared at about 10 days. The CNS was markedly deficient in myelin, but the peripheral nerves were myelinated. The phenotype was similar to that resulting from *jp* in many respects, but in view of inconclusive tests for allelism, the authors suggested that the 2 loci were different. Biochemically, *msd*/Y mice were found to resemble *jp*/Y in several respects (Brenkert et al. 1972; Kandutsch and Saucier 1972) and to differ in others (Sarlieve et al. 1974). Eicher and Hoppe (1973) proved *jp* and *msd* to be alleles by creating an experimental chimeric male that transmitted *jp*. This male produced abnormal female offspring when crossed to *jp*/+ and *msd*/+ females. The symbol *jp^{msd}* was recommended. After a detailed morphological study of *jp*/Y animals, Billings-Gagliardi et al. (1980a) concluded that, although qualitatively similar, the two mutants are quantitatively different in that jp^{msd}/Y animals consistently have about twice as much myelin as jp/Y animals in the 3 regions of the brain that were studied. When placed on the same genetic background, the disease produced by jp continued to be more severe than that produced by jp^{msd} (Wolf et al. 1983). The expression of the 2 mutants in vitro is qualitatively similar (Billings-Gagliardi et al. 1980b). Gardinier and Macklin (1988) found that affected mice are capable of synthesizing normal PLP and DM-20. However, PLP mRNA levels are drastically reduced and the normal developmental pattern of expression for both proteins is altered. These results suggest that the jp^{msd} mutation is a sequence variation residing within the promoter region or a regulatory element outside the PLP coding region.

Rat, myelin deficiency, *md*

The first mutants with this disorder were observed in 1977 among a colony of Wistar rats in Albany, New York (Csiza and deLahunta 1979). The first recognizable sign of the disease in affected males is a head tremor at 12–15 days of age. The tremors become generalized within a few days, but disappear when the pups rest. In the later stages, from 17 to 21 days, the slightest disturbance precipitates a generalized seizure. Affected pups normally die at 24–28 days, but life can be prolonged by an anticonvulsant (Dentinger et al. 1982). At autopsy, the spinal cord appears gray rather than white. Microscopically, the CNS lacks myelin formation; there is no evidence of myelin destruction. The peripheral nerves are myelinated. Ultrastructurally, no normal myelin sheath can be detected at any level of the CNS (Barron et al. 1980). The oligodendroctyes are affected primarily; they fail to form compact myelin and contain numerous electron-lucent vacuoles, apparently derived from the endoplasmic reticulum, or large numbers of membrane-bound, homogeneous lipid inclusions; axonal abnormalities are present in 17- to 20-day-old rats. The PNS appears normal. Additional detailed pathological changes were described by Dentinger et al. (1982, 1985), Barron et al. (1987), and Duncan et al. (1987b). Csiza (1982) mentions 5 affected females in which the disease was "clinically, histologically, electron-microscopically and biochemically indistinguishable" from that in males. Duncan et al. (1987a) found that heterozygotes show myelin mosaicism of the optic nerve and spinal cord. Abnormal oligodendrocytes with distended rough endoplasmic reticulum are found in the abnormal myelin patches. All myelin proteins are severely reduced in affected animals, but PLP is reduced more than the others and cannot be detected by a very sensitive technique (Yanagisawa et al. 1986). Blot hybridization analysis of RNA and DNA from affected rats indicates that the responsible mutation involves *Plp* and suggests that the error is similar to that in *jp* (see above) (Naismith et al. 1987).

Barron KD, Dentinger MP, Csiza CK: 1980. Ultrastructural observations on myelin deficiency (*md*), a dysmyelinating neurologic mutant in the Wistar rat. In: Baumann N (ed), *Neurological Mutations Affecting Myelination*. Amsterdam: Elsevier/North Holland Biomedical Press, pp 99–106.

Barron KD, Dentinger MP, Csiza CK, Keegan SM, Mankes R: 1987. Abnormalities of central axons in a dysmyelinative rat mutant. *Exp Mol Pathol* 47: 125–142.

Bartlett WP, Skoff RP: 1986. Expression of the jimpy gene in the spinal cords of heterozygous female mice. 1. An early myelin deficit followed by compensation. *J Neurosci* 6: 2802–2812.

Benjamins JA, Skoff RP, Beyer K: 1984. Biochemical expression of mosaicism in female mice heterozygous for the jimpy gene. *J Neurochem* 42: 487–492.

Benjamins JA, Studzinski DM, Skoff RP: 1986. Biochemical correlates of myelination in brain and spinal cord of mice heterozygous for the jimpy gene. *J Neurochem* 47: 1857–1863.

Billings-Gagliardi S, Adcock LH, Wolf MK: 1980a. Hypomyelinated mutant mice: description of jp^{msd} and comparion with jp and qk on their present genetic backgrounds. *Brain Res* 194: 325–338.

Billings-Gagliardi S, Adcock LH, Schwing GB, Wolf MK: 1980b. Hypomyelinated mutant mice. II. Myelination in vitro. *Brain Res* 200: 135–150.

Brenkert A, Arora RC, Radin NS, Meier H, MacPike AD: 1972. Cerebroside synthesis and hydrolysis in a neurological mutant mouse (*msd*). *Brain Res* 36: 195–202.

Campagnoni AT, Macklin WB: 1988. Cellular and molecular aspects of myelin protein gene expression. *Mol Neurobiol* 2: 41–89.

Csiza CK: 1982. Lipid class analysis of the central nervous system of myelin-deficient Wistar rats. *J Lip Res* 23: 720–725.

Csiza CK, deLahunta A: 1979. Myelin deficiency (md): a neurological mutant in the Wistar rat. *Am J Pathol* 95: 215–224.

Dautigny A, Alliel PM, d'Auriol L, Pham-Dinh D, Nussbaum J-L, Galibert F, Jollès P: 1985. Molecular cloning and nucleotide sequence of a cDNA clone coding for rat brain myelin proteolipid. *FEBS Lett* 188: 33–36.

Dautigny A, Mattei M-G, Morello D, Alliel PM, Pham-Dinh D, Amar L, Arnaud D, Simon D, Mattei J-F, Guenet J-L, Jollès P, Avner P: 1986. The structural gene coding for myelin-associated proteolipid protein is mutated in jimpy mice. *Nature* 321: 867–869.

Dentinger MP, Barron KD, Csiza CK: 1982. Ultrastructure of the central nervous system in a myelin deficient rat. *J Neurocytol* 11: 671–691.

Dentinger MP, Barron KD, Csiza CK: 1985. Glial and axonal development in optic nerve of myelin deficient rat mutant. *Brain Res* 344: 255–266.

Diehl H-J, Schaich M, Budzinski R-M, Stoffel W: 1986. Individual exons encode the integral membrane domains of human myelin proteolipid protein. *Proc Natl Acad Sci USA* 83: 9807–9811.

Duncan ID, Griffiths IR, Munz M: 1983. "Shaking pups": a disorder of central myelination in the spaniel dog. III. Quantitative aspects of glia and myelin in the spinal cord and optic nerve. *Neuropathol Appl Neurobiol* 9: 355–368.

Duncan ID, Hammang JP, Jackson KF: 1987a. Myelin mosaicism in female heterozygotes of the canine shaking pup and myelin-deficient rat mutants. *Brain Res* 402: 168–172.

Duncan ID, Hammang JP, Trapp BD: 1987b. Abnormal compact myelin in the myelin-deficient rat: absence of proteolipid protein correlates with a defect in the intraperiod line. *Proc Natl Acad Sci USA* 84: 6287–6291.

Eicher EM, Hoppe PC: 1973. Use of chimeras to transmit lethal genes in the mouse and to demonstrate allelism of the two X-linked male lethal genes *jp* and *msd*. *J Exp Zool* 183: 181–184.

Gardinier MV, Macklin WB: 1988. Myelin proteolipid protein gene expression in jimpy and jimpy[msd] mice. *J Neurochem* 51:360–369.

Gardinier MV, Macklin WB, Diniak AJ, Deininger PL: 1986. Characterization of myelin proteolipid mRNAs in normal and jimpy mice. *Mol Cell Biol* 6: 3755–3762.

Griffiths IR, Duncan ID, McCulloch M, Harvey MJA: 1981a. Shaking pups: a disorder of central myelination in the spaniel dog. I. Clinical, genetic and light-microscopical observations. *J Neurol Sci* 50: 423–433.

Griffiths IR, Duncan ID, McCulloch M: 1981b. Shaking pups: a disorder of central myelination in the spaniel dog. II. Ultrastructural observations on the white matter of the cervical spinal cord. *J Neurocytol* 10: 847–858.

Hatfield JS, Skoff RP: 1982. GFAP immunoreactivity reveals astrogliosis in females heterozygous for jimpy. *Brain Res* 250: 123–131.

Hudson LD, Berndt JA, Puckett C, Kozak CA, Lazzarini RA: 1987. Aberrant splicing of proteolipid protein in RNA in the dysmyelinating jimpy mutant mouse. *Proc Natl Acad Sci USA* 84: 1454–1458.

Hogan EL, Greenfield S: 1984. Animal models of genetic disorders of myelin. In: Morell P (ed), *Myelin*, 2nd ed. New York: Plenum Press, pp 491–499.

Ikenaka K, Furuichi T, Iwasaki Y, Moriguchi A, Okano H, Mikoshiba K: 1988. Myelin proteolipid protein gene structure and its regulation of expression in normal and *jimpy* mutant mice. *J Mol Biol* 199: 587–596.

Jollès J, Nussbaum J-L, Jollès P: 1983. Enzymic and chemical fragmentation of the apoprotein of the major rat brain myelin proteolipid: sequence data. *Biochim Biophys Acta* 742: 33–38.

Kandutsch AA, Saucier SE: 1972. Sterol and fatty acid synthesis in developing brains of three myelin-deficient mouse mutants. *Biochim Biophys Acta* 260: 26–34.

Kerner A-L, Carson JH: 1984. Effect of the jimpy mutation on expression of myelin proteins in heterozygous and hemizygous mouse brain. *J Neurochem* 43: 1706–1715.

Knapp PE, Skoff RP, Redstone DW: 1986. Oligodendroglial cell death in jimpy mice: an explantion for the myelin deficit. *J Neurosci* 6: 2813–2822.

Koeppen AH, Ronca NA, Greenfield EA, Hans MB: 1987. Defective biosynthesis of proteolipid protein in Pelizaeus–Merzbacher disease. *Ann Neurol* 21: 159–170.

Kronquist KE, Crandall BF, Macklin WB, Campagnoni AT: 1987. Expression of myelin proteins in the developing human spinal cord: cloning and sequencing of human proteolipid protein cDNA. *J Neurosci Res* 18: 395–401.

Lees MB, Brostoff SW: 1984. Proteins of myelin. In: Morell P (ed), *Myelin*, 2nd ed. New York: Plenum Press, pp 197–224.

Lerner P, Campagnoni AT, Sampugna J: 1974. Proteolipid in the developing brains of normal mice and myelin deficient mutants. *J Neurochem* 22: 163–170.

Lyon G, Goffinet A: 1980. Genetics and pathology of dysmyelinating disorders of the central nervous system. Comparison to animal models. In: Baumann N (ed), *Neurological Mutations Affecting Myelination*. Amsterdam: Elsevier/North Holland Biomedical Press, p 40.

Macklin WB, Campagnoni CW, Deininger PL, Gardinier MV: 1987. Structure and expression of the mouse myelin proteolipid protein gene. *J Neurosci Res* 18: 383–394.

Mattei MG, Alliel PM, Dautigny A, Passage E, Pham-Dinh D, Mattei JF, Jollès P: 1986. The gene encoding for the major brain proteolipid (PLP) maps on the q-22 band of the human X chromosome. *Hum Genet* 72: 352–353.

Matthieu J-M, Quarles RH, Webster H de F, Hogan EL, Brady RO: 1974. Characterization of the fraction obtained from the CNS of jimpy mice by a procedure for myelin isolation. *J Neurochem* 23: 517–523.

Meier C, Bischoff A: 1975. Oligodendroglial cell development in jimpy mice and controls. An electron-microscopic study in the optic nerve. *J Neurol Sci* 26: 517–528.

Meier H, MacPike AD: 1970. A neurological mutation (msd) of the mouse causing a deficiency of myelin synthesis. *Exp Brain Res* 10: 512–525.

Milner RJ, Lai C, Nave K-A, Lenoir D, Ogata J, Sutcliffe JG: 1985. Nucleotide sequences of two mRNAs for rat brain myelin proteolipid protein. *Cell* 42: 931–939.

Morello D, Dautigny A, Pham-Dinh D, Jollès P: 1986. Myelin proteolipid protein (PLP and DM-20) transcripts are deleted in *jimpy* mutant mice. *EMBO J* 5: 3489–3493.

Moriguchi A, Ikenaka K, Furuichi T, Okano H, Iwasaki Y, Mikoshiba K: 1987. The fifth exon of the myelin protein-coding gene is not utilized in the brain of *jimpy* mutant mice. *Gene* 55: 333–337.

Naismith AL, Simons R, Alon N, Fahim S, Csiza CK, Riordan JR: 1987. Structure and expression of the myelin PLP gene. (Abstr) *J Neurochem* 48 (Suppl): S32 only.

Nave K-A, Lai C, Bloom FE, Milner RJ: 1986. Jimpy mutant mouse: a 74-base deletion in the mRNA for myelin proteolipid protein and evidence for a primary defect in RNA splicing. *Proc Natl Acad Sci USA* 83: 9264–9268.

Nave K-A, Lai C, Bloom FE, Milner RJ: 1987a. Splice site selection in the proteolipid protein (PLP) gene transcript and primary structure of the DM-20 protein of central nervous system myelin. *Proc Natl Acad Sci USA* 84: 5665–5669.

Nave K-A, Bloom FE, Milner RJ: 1987b. A single nucleotide difference in the gene for myelin proteolipid protein defines the *jimpy* mutation in mouse. *J Neurochem* 49: 1873–1877.

Nussbaum JL, Mandel P: 1973. Brain proteolipids in neurological mutant mice. *Brain Res* 61: 295–310.

Phillips RJS: 1954. *Jimpy*, a totally sex-linked gene in the house mouse. *Z Indukt Abstammungs-Vererbungsl* 86: 322–326.

Privat A, Robain C, Mandel P: 1972. Aspects ultrastructuraux du corps calleux chez la souris Jimpy. *Acta Neuropathol (Berl)* 21: 282–295.

Rosenfeld J, Friedrich VL Jr: 1986. Oligodendrocyte production and myelin recovery in heterozygous jimpy mice: an autoradiographic study. *Int J Dev Neurosci* 4: 179–187.

Roussel G, Neskovic NM, Trifilieff E, Artault J-C, Nussbaum J-L: 1987. Arrest of proteolipid transport through the Golgi apparatus in Jimpy brain. *J Neurocytol* 16: 195–204.

Sarlieve LL, Neskovic NM, Rebel G, Mandel P: 1974. PAPS–cerebroside sulphotransferase activity in developing brain of a neurological mutant of mouse (*msd*). *Exp Brain Res* 19: 158–165.

Seitelberger F: 1954. Die Pelizaeus–Merzbachersche Krankheit. *Wien Z Nervenheilk* 9: 228–289.

Sidman RL, Dickie MM, Appel SH: 1964. Mutant mice (quaking and jimpy) with deficient myelination in the central nervous system. *Science* 144: 309–311.

Sidman RL, Green MC, Appel SH: 1965. *Catalog of the Neurological Mutants of the Mouse.* Cambridge, MA: Harvard Univ Press, pp 27–29.

Skoff R, Nowicki Montgomery I: 1981. Expression of mosaicism in females heterozygous for jimpy. *Brain Res* 212: 175–181.

Sorg BJA, Agrawal D, Agrawal HC, Campagnoni AT: 1986. Expression of myelin proteolipid protein and basic protein in normal and dysmyelinating mutant mice. *J Neurochem* 46: 379–387.

Stoffel W, Giersiefen H, Hillen H, Schroeder W, Tunggal B: 1985. Amino-acid sequence of human and bovine brain myelin proteolipid protein (lipophilin) is completely conserved. *Biol Chem Hoppe–Seyler* 366: 627–635.

Torii J, Adachi M, Volk BW: 1971. Histochemical and ultrastructural studies of inherited leukodystrophy in mice. *J Neuropathol Exp Neurol* 30: 278–289.

Tsuji S: 1985. Successful treatment with forphenicinol, an immunomodifier, to prevent the defective myelination in jimpy mouse. (Japanese, English summary) *Proc Jpn Assoc Anim Models Hum Dis* 1: 13–17.

Willard HF, Riordan JR: 1985. Assignment of the gene for myelin proteolipid protein to the X chromosome: implications for X-linked myelin disorders. *Science* 230: 940–942.

Wolf MK, Holden AB: 1969. Tissue culture analysis of the inherited defect of central nervous system myelination in jimpy mice. *J Neuropathol Exp Neurol* 28: 195–213.

Wolf MK, Kardon GB, Adcock LH, Billings-Gagliardi S: 1983. Hypomyelinated mutant mice. V. Relationship between *jp* and *jp^{msd}* re-examined on identical genetic backgrounds. *Brain Res* 271: 121–129.

Yanagisawa K, Quarles RH: 1986. Jimpy mice: quantitation of myelin-associated glycoprotein and other proteins. *J Neurochem* 47: 322–325.

Yanagisawa K, Duncan ID, Hammang JP, Quarles RH: 1986. Myelin-deficient rat: analysis of myelin proteins. *J Neurochem* 47: 1901–1907.

Yanagisawa K, Moller JR, Duncan ID, Quarles RH: 1987. Disproportional expression proteolipid protein and DM-20 in the X-linked, dysmyelinating shaking pup mutant. *J Neurochem* 49: 1912–1917.

31217 QRS INTERVAL [NK]

Horse

In Australia, electrocardiography has been used as a noninvasive technique to determine heart size in racehorses. The procedure involves careful measurement of the QRS interval in each of the standard limb leads, and calculation of the mean QRS duration in milliseconds; this value is referred to as the "heart score" and is highly positively correlated with heart weight and racing performance (Steel et al. 1977). In studies to assess the genetic component in the heart score, Steel and his colleagues found evidence of the contribution of the X chromosome in Thoroughbreds but not in Standardbreds.

Steel JD, Beilharz RG, Stewart GA, Goddard M: 1977. The inheritance of heart score in racehorses. *Aust Vet J* 53: 306–309.

*31236 RAUSCHER LEUKEMIA VIRUS SUSCEPTIBILITY-3 [NK]

Mouse, *Rv-3*

Rauscher leukemia virus (RLV), like Friend leukemia virus but unlike other murine viruses, produces a rapid erythroleukemia in susceptible adult mice (Rauscher 1962).

RLV induces macroscopic foci of primitive erythroid cells in the spleen (Pluznik and Sachs 1964). Two autosomal loci controlling susceptibility to RLV are known (Toth et al. 1973). Heller and Pluznik (1984) described an X-linked gene that contributes to susceptibility to spleen focus formation by RLV; hemizygous male offspring of crosses between resistant-strain females (C57BL/6J) and susceptible-strain males (CBA/Lac) showed significantly lower numbers of foci/spleen than did their female littermates and the offspring of both sexes of the reciprocal cross.

Heller E, Pluznik DH: 1984. Chromosomal assignment of two murine genes controlling susceptibility to spleen focus formation by Rauscher leukemia virus. *Exp Hematol* 12: 645–649.

Pluznik DH, Sachs L: 1964. Quantitation of a murine leukemia virus with a spleen colony assay. *J Natl Cancer Inst* 33: 535–546.

Toth FD, Vaczi L, Balogh M: 1973. Inheritance of susceptibility and resistance to Rauscher leukemia virus. *Acta Microbiol Hung* 20: 183–189.

Rauscher FJ JR: 1962. A virus-induced disease of mice characterized by erythrocytopoiesis and lymphoid leukemia. *J Natl Cancer Inst* 29: 515–543.

*31276 RIBONUCLEIC ACID, 7S (7SRNA; RN7S) [NK]

7SRNA, a highly conserved abundant cytoplasmic species, plays a critical role in moving secretory proteins from the ribosome to the endoplasmic reticulum.

Mouse, Rn7s-X

Taylor et al. (1984) and Taylor and Rowe (1984, 1985) used a mouse embryo cDNA plasmid clone, which contains part of a 7SRNA gene sequence, to detect restriction fragment length variants among inbred strains. 7SRNA related sequences are widely dispersed in the mouse genome, and 1 is X-linked (Taylor and Rowe 1984).

Taylor BA, Rowe L: 1984. Private communication. *Mouse News Lett* 71: 31 only.

Taylor BA, Rowe L: 1985. Private communication. *Mouse News Lett* 72: 110 only.

Taylor BA, Rowe L, Gibbons K: 1984. Private communication. *Mouse News Lett* 70: 84 only.

*31288 SEGREGATION REVERSAL [NK]

Mouse, *Segr*

After observing that the normal X chromosome and the X chromosomes from 2 methylcholanthrene-induced sarcoma lines are able to switch the chromosome segregation of mouse–Chinese hamster somatic cell hybrids, Pravtcheva and Ruddle (1983a,b) concluded that this reversal is mediated by factors on the X chromosome.

Pravtcheva DD, Ruddle FH: 1983a. X chromosome-induced reversion of chromosome segregation in mouse/Chinese hamster somatic cell hybrids: cellular recognition of native and foreign X chromosomes. *Exp Cell Res* 146: 401–416.

Pravtcheva DD, Ruddle RH: 1983b. Normal X chromosome induced reversion in the direction of chromosome segregation in mouse–Chinese hamster somatic cell hybrids. *Exp Cell Res* 148: 265–272.

31336 SPLAYLEG [NK]

Pig

Lax (1971) described this disorder in which newborn pigs, of a Moravian Czech White and Landrace crossbred herd, were unable to stand; the fore- and hindlimbs were splayed forward and sideways. The condition, first presented as a "genetic myopathy" at a meeting in Brno in 1966, disappeared within 18 days in those animals that survived injuries caused by the abnormal posture. Vogt et al. (1984) ascribe death to inability to suckle and accidental crushing by the sow; survivors usually recovered within a week after delivery. Although Lax suggested an X-linked dominant model of inheritance "with varying degrees of penetrance," the data presented do not support this view. An identical phenotype was described as myofibrillar hypoplasia (Thurley et al. 1967) and spraddle-leg (Olson and Prange 1968). Histological and biochemical studies suggest that the myofibrillar hypoplastic muscle is simply immature muscle (Patterson et al. 1969; Thurley and Done 1969). Vogt et al. (1984) carried out extensive studies on the University of Missouri–Columbia herd and found no evidence of a simple Mendelian mode of inheritance. Their conclusion that the trait has a multifactorial etiology is consistent with earlier observations by Dobson (1968, 1971) and Thurley et al. (1967). Landrace boars appear to produce consistently a greater proportion of affected piglets than do boars of other breeds (Dobson 1968; Lax 1971; Vogt et al. 1984). The disorder was reviewed extensively by Ward (1978).

Dobson KJ: 1968: Congenital splayleg of piglets. *Aust Vet J* 44: 26–28.
Dobson KJ: 1971. Failure of choline and methionine to prevent splayleg in piglets. *Aust Vet J* 47: 587–590.
Lax T: 1971. Hereditary splayleg in pigs. *J Hered* 62: 250–252.
Olson LRD, Prange JF: 1968. Spraddle-legged baby pigs. *Vet Med Small Anim Clin* 63: 714 only.
Patterson DSP, Sweasey D, Allen WM, Berrett S, Thurley DC: 1969. The chemical composition of neonatal piglet muscle and some observations on the biochemistry of myofibrillar hypoplasia occurring in otherwise normal litters. *Zentralbl Veterinarmed [A]*16: 741–753.
Thurley DC, Done JT: 1969. The histology of myofibrillar hypoplasia of newborn pigs. *Zentralbl Veterinarmed [A]* 16: 732–740.
Thurley DC, Gilbert FR, Done JT: 1967. Congenital splayleg of piglets: myofibrillar hypoplasia. *Vet Rec* 80: 302.–304.
Vogt DW, Gipson TA, Akremi B, Dover S, Ellersieck MR: 1984. Associations of sire, breed, birth weight, and sex in pigs with congential splayleg. *Am J Vet Res* 45: 2408–2409.
Ward PS: 1978. The splayleg syndrome in new-born pigs: a review. *Vet Bull* 48: 279–295; 381–399.

*31341 STEROID SULFATASE, X-LINKED (STERYL SULFATASE; ARYLSULFATASE C; STERYL SULFATE SULFOHYDROLASE; STS; EC 3.1.6.2) [30810]

This is probably the structural locus for this microsomal enzyme that is involved in the metabolic activation of a variety of steroid compounds. In man, a mutation at this locus causes an enzyme deficiency demonstrable in cultured skin fibroblasts of males with an X-linked form of ichthyosis (MIM 30810). The locus is on the short arm, and like Xg^a, to which it is linked, it is not normally inactivated.

Marsupials

Cooper et al. (1984) and Dawson and Graves (1986) were unable to detect the locus on the X chromosome of 2 superfamilies of Australian marsupials, the Phalangeroidae (Kangaroos) and the Dasyuroidae.

Mouse and other rodents, *Sts*

Ropers et al. (1982) and Ropers and Wiberg (1982) reported that STS levels in the wood lemming (*Myopus schisticolor*) are linearly correlated with the number of X chromosomes. This finding indicates that in this rodent species the locus is X-linked and not inactivated. Ropers et al. (1982) also reported that no complementation occurred in clones of hybrids of STS-negative murine cells and STS-deficient human fibroblasts, and that "circumstantial evidence" indicated that the STS locus was X-linked in *Microtus agrestis* L. Gartler and Rivest (1983) found that STS levels are approximately twice as high in XX oocytes as in XO oocytes of *M. musculus*, indicating that the locus is X-linked in this species; however, assays of STS indicated that dosage compensation occurs for the gene in the kidney tissue of XX and XO animals. Lam et al. (1983) and Crocker and Craig (1983) also showed that the murine locus is inactivated. Balazs et al. (1982) described what appeared to be an autosomally controlled deficiency in C3H/An mice. However, Keitges et al. (1985), by crossing STS-normal XO females with C3ZH/An STS-deficient males, demonstrated that the STS deficiency is controlled by an X-linked gene. The 2 sets of results, one suggesting autosomal, the other X-linked control, indicate the existence of an STS allele on the Y chromosome that undergoes obligatory recombination. Keitges et al. (1987) used linkage studies with *Sxr* (Evans et al. 1982; Singh and Jones 1982) to provide direct evidence for a functional *Sts* gene pair in the X-Y pairing region. They demonstrated that at least 1 obligatory crossover occurs between the X and Y chromosomes; the crossing-over does not occur at a single point but over a region. The linkage of *Sts* to *Sxr* was confirmed by Nagamine et al. (1987). Keitges and Gartler (1986) investigated dosage compensation at the locus by measuring STS activity in various tissues of XX, XO, and XY offspring from crosses of XO females and XY males. Both X chromosomes and the Y-linked allele expressed STS. In kidney and ear punch samples, there was no evidence that *Sts* is inactivated; however, in spleen tissue, dosage compensation for *Sts* expression may not be complete. Cattanach and Crocker (1986) showed that the locus is X-linked and provided support for the prediction of Keitges et al. (1985) that it should be located at the distal, X-Y pairing, end of the chromosome. STS levels are significantly higher in male than in female fibroblasts (1.6:1) of the root vole (*Microtus oeconomus*), which belongs to the same family as the wood lemming (Wiberg and Fredga 1987). This finding is strikingly different from that in the wood lemming (see above). Wiberg and Fredga discuss possible explantions for this difference. Craig and Tolley (1986) succinctly reviewed this fascinating locus.

Balazs I, Purrello M, Rocchi M, Rinaldi A, Siniscalco M: 1982. Is the gene for steroid sulfatase X-linked in man? An appraisal of data from humans, mice, and their hybrids. (Abstr) *Cytogenet Cell Genet* 32: 251–252.

Cattanach BM, Crocker M: 1986. X chromosomal location of *Sts*. *Mouse News Lett* 74: 94–95.

Cooper DW, McAllan BM, Donald JA, Dawson G, Dobrovic A, Graves JAM: 1984. Steroid sulphatase is not detected on the X chromosome of Australian marsupials. (Abstr) *Cytogenet Cell Genet* 37: 439 only.

Craig W, Tolley E: 1986. Steroid sulphatase and the conservation of mammalian X chromosomes. *Trends Genet* 2: 201–203.

Crocker M, Craig I: 1983. Variation in regulation of steroid sulphatase locus in mammals. *Nature* 303: 721–722.

Dawson GW, Graves JAM: 1986. Gene mapping in marsupials and monotremes. III. Assignment of four genes to the X chromosome of the wallaroo and the euro (*Macropus robustus*). *Cytogenet Cell Genet* 42: 80–84.

Evans EP, Burtenshaw MD, Cattanach BM: 1982. Meiotic crossing over between the X and Y chromosomes of male mice carrying the sex-reversing (*Sxr*) factor. *Nature* 300: 443–445.

Gartler SM, Rivest M: 1983. Evidence for X-linkage of steroid sulfatase in the mouse: steroid sulfatase levels in oocytes of *XX* and *XO* mice. *Genetics* 10: 137–141.

Keitges EA, Gartler SM: 1986. Dosage of the *Sts* gene in the mouse. *Am J Hum Genet* 39: 470–476.

Keitges E, Rivest M, Siniscalco M, Gartler SM: 1985. X-linkage of steroid sulphatase in the mouse is evidence for a function Y-linked allele. *Nature* 315: 226–227.

Keitges EA, Schorderet DF, Gartler SM: 1987. Linkage of the steroid sulfatase gene to the *sex-reversed* mutation in the mouse. *Genetics* 116: 465–468.

Lam STS, Polani PE, Fensom AH: 1983. Steroid sulphatase in the mouse. *Genet Res* 41: 299–302.

Nagamine CM, Michot J-L, Roberts C, Guenet J-L, Bishop CE: 1987. Linkage of the murine steroid sulfatase locus, *Sts*, to sex reversed, *Sxr*: a genetic and molecular analysis. *Nucleic Acids Res* 15: 9227–9238.

Ropers H-H, Wiberg U: 1982. Evidence for X-linkage and non-inactivation of steroid sulphatase locus in wood lemming. *Nature* 296: 766–767.

Ropers H-H, Wiberg V, Migl B, Sperling K, Fredga K: 1982. X-linkage of STS in rodent
    species: evolutionary conservation of non-inactivated X chromosome segment. (Abstr)
    Cytogenet Cell Genet 32: 312 only.
Singh L, Jones KW: 1982. Sex reversal in the mouse (Mus musculus) is caused by a recurrent
    nonreciprocal crossover involving the X and an aberrrant Y chromosome. *Cell* 28:
    205–216.
Wiberg UH, Fredga K: 1987. Steroid sulfatase levels are higher in males than in females of
    the root vole (*Microtus oeconomus*). Yet another rodent with an active Y-linked allele?
    Hum Genet 77: 6–11.

*31343 STREAKED HAIRLESSNESS [?30560, ?30830]

Cattle

Eldridge and Atkeson (1953) described this abnormality of hair coat in a herd of
Holstein–Friesian cattle in Kansas. In affected females, approximately perpendicular,
irregularly narrow hairless streaks occurred on various parts of the body; no other
anomalies were observed. There were no affected males, and analysis of the sex ratio
and calving interval in the herd suggested that hemizygotes died early in utero.
Although McKusick (1986) suggests homology with focal dermal hypoplasia (MIM
30650) or incontinentia pigmenti (MIM 30830) in man, and striated (31349) in the
mouse, the evidence is not strong. I do not know if this mutation is still being
maintained.

Eldridge FE, Atkeson FW: 1953. Streaked hairlessness in Holstein–Friesian cattle: a
    sex-linked, lethal character. *J Hered* 44: 265–271.
McKusick VA: 1986. *Mendelian Inheritance in Man: Catalogs of Autosomal Dominant,
    Autosomal Recessive, and X-Linked Phenotypes*, 7th ed. Baltimore: Johns Hopkins Univ.
    Press.

*31344 SYNAPSIN 1 (SYN1)

Synapsin 1 is a neuron-specific phosphoprotein associated with the membranes of
synaptic vesicles throughout the nervous system (Greengard 1981; De Camilli et al.
1983). In the developing cerebella of the rat and the chick, the appearance of
immunohistochemically detectable synapsin 1 correlates temporally and topograph-
ically with the appearance of synapses (De Camilli et al. 1983). Yang-Feng et al.
(1986) used synapsin 1 cDNAs cloned from a rat brain cDNA library by Kilimann
and DeGennaro (1985) to assign the gene to the proximal region of the short arm of
the human X chromosome (Xp 21.1→ cen) by in situ hybridization. Yang-Feng et al.
hypothesize that mutations at this locus could give rise to disorders with primary
neuronal degeneration, such as the Rett syndrome.

Mouse, *Syn-1*

Yang-Fenn et al. (1986) used somatic cell hybridization to map the locus to the XA1→XA4 region, which also contains *spf* (31125); the 2 loci are also closely linked in man.

De Camilli P, Cameron R, Greengard P: 1983. Synapsin I (protein I), a nerve terminal-specific phosphoprotein. I. Its general distribution in synapses of the central and peripheral nervous system demonstrated by immunofluorescence in frozen and plastic sections. *J Cell Biol* 96: 1337–1354.

Greengard P: 1981. Intracellular signals in the brain. *Harvey Lectures, Ser 75*: 277–331 (pp 287–307).

Kilimann MW, DeGennaro LJ: 1986. Molecular cloning of cDNAs for the nerve-cell specific phosphoprotein, synapsin I. *EMBO J* 4: 1997–2002.

Yang-Feng TL, DeGennaro LJ, Francke U: 1986. Genes for synapsin I, a neuronal phosphoprotein, map to conserved regions of human and murine X chromosomes. *Proc Natl Acad Sci USA* 83: 8679–8683.

31347 SURFACE ANTIGEN MIC2 (MIC2; MONOCLONAL ANTIBODY 12E7: MIC2X)

Chimpanzee and gorilla

12E7 is a human cell-surface antigen defined by a unique monoclonal antibody. The expression of the antigen on nucleated cells is independently controlled by 2 genes: MIC2X and MIC2Y (Goodfellow 1983). The former is located at the tip of the short arm of the X (Xp22,32-Xpter), and the latter in the euchromatic part of the Y (Yp11.2-Ypter) closely linked to *TDF* (Buckle et al. 1985; Goodfellow et al. 1986). Dermal fibroblasts and red blood cells from gorillas and chimpanzees express the 12E7 antigen, but there is no evidence for quantitative polymorphism (Shaw et al. 1985). Orangutan fibroblasts; the red blood cells of gibbons (12), baboons (4), goats (3), sheep (3), cows (3), and deer (2); and cultured cells from a wide variety of other mammals do not express the 12E7 antigen (Goodfellow 1983; Shaw et al. 1985). The locus has not been mapped in chimpanzees and gorillas.

Buckle V, Mondello C, Darling S, Craig IW, Goodfellow PN: 1985. Homologous expressed genes in the human sex chromosome pairing region. *Nature* 317: 739–741.

Goodfellow P: 1983. Expression of the 12R7 antigen is controlled independently by genes on the human X and Y chromosomes. *Differentiation* 23 (Suppl): S35–S39.

Goodfellow PJ, Darling SM, Thomas NS, Goodfellow PN: 1986. A pseudoautosomal gene in man. *Science* 234: 740–743.

Shaw MA, Tippett P, Delhanty JDA, Andrews M, Goodfellow P: 1985. Expression of Xg and the 12E7 antigen in primates. *J Immunogenet* 12: 115–118.

*31351 STRIATED [?30560, ?30830]

Mouse, *Str*

This mutation was first described by Phillips (1963). The original mutant was a female born to an X-irradiated male. Although its phenotype was similar to a tabby heterozygote (30510), the new mutant differed from it in that the striations were less well marked and often not distinguishable until 16–18 days; Lyon (1963) found that the black-striped regions of the coat of agouti animals results from a shortening of hairs, not from the lack of zigzag hairs observed in tabby females. The short hairs are grossly abnormal, and the long ones have minor irregularities (Grüneberg 1966). The gene is only about 80% penetrant, and hemizygotes die at between 11½ and 13 days of gestation. The original linkage relations reported by Phillips (1963) were incorrect (Lyon 1966); the correct order is *Bn-Str-Ta*. McKusick (1986) suggests homology with focal dermal hypoplasia (MIM 30560) or incontinentia pigmenti (MIM 30830) in man and streaked hairlessness (31343) in the cow, but the evidence is not convincing; Yang-Feng et al. (1986: 8683) point out that, considering relative map positions, tattered (31352) is more likely to be homologous with incontinentia pigmenti.

Grüneberg H: 1966. More about the tabby mouse and the Lyon hypothesis. *J Embryol Exp Morphol* 16: 569–590.

Lyon MF: 1963. Attempts to test the inactive-X theory of dosage compensation in mammals. *Genet Res* 4: 93–103.

Lyon MF: 1966. Order of loci on the X-chromosome of the mouse. *Genet Res* 7: 130–133.

McKusick VA: 1986. *Mendelian Inheritance in Man: Catalogs of Autosomal Dominant, Autosomal Recessive, and X-Linked Phenotypes*, 7th ed. Baltimore: Johns Hopkins Univ Press.

Phillips RJS: 1963. *Striated*, a new sex-linked gene in the house mouse. *Genet Res* 4: 151–153.

Yang-Feng TL, Degennaro LJ, Francke U: 1986. Genes for synapsin I, a neuronal phosphoprotein, map to conserved regions of human and murine X chromosomes. *Proc Natl Acad Sci USA* 83: 8679–8683.

*31352 TATTERED [?30830]

Mouse, *Td*

This mutation was first detected in a female offspring of an X-irradiated male (Cattanach 1982a). It is a prenatal lethal in males. Heterozygotes have reduced viability (60–80% of normal), and apparently die both pre- and postnatally. The defect can be recognized at 5–6 days when scarring of the skin occurs in patches or streaks over the whole body including the tail, legs, and feet; hair growth is restricted to the unscarred regions. In moderately affected adults, the coat has a striped appearance similar to that in tabby (30510) or bare patches (30295), but in more severely affected animals, the coat has a tattered appearance caused by many small irregular bald areas.

At weaning, heterozygotes are smaller and weigh less than their normal litter mates. The locus is at the proximal end of the chromosome very close to *spf* (31125) and *sf* (30820) (Cattanach 1982b). On the basis of relative map positions, Yang-Feng et al. (1986: 8683) suggest that tattered might be homologous with incontinentia pigmenti (MIM 30830).

Cattanach BM: 1982a. Private communication. *Mouse News Lett* 66: 61–62.
Cattanach BM: 1982b. Private communication. *Mouse News Lett* 67: 19 only.
Yang-Feng TL, DeGennaro LJ, Francke U: 1986. Genes for synapsin I, a neuronal phosphoprotein, map to conserved regions of human and murine X chromosomes. *Proc Natl Acad Sci USA* 83: 8679–8683.

*31353 TAWNY [NK]

Syrian (golden) hamster, *T*

This mutation produces paleness of the normal golden hue of the wild-type hamster coat (Magalhaes 1954). X-linkage was demonstrated by Peterson et al. (1977).

Magalhaes H: 1954. Cream and tawny, coat color mutations in golden hamster, *Mesocricetus auratus*. (Abstr) Anat Rec 120: 752 only.
Peterson JS, Kirtland-Coffey A, Yoon CH: 1977. Tawny: a sex-linked, incompletely dominant trait in Syrian hamsters. *J Hered* 68: 131–132.

*31354 TEAR PROTEIN SYSTEM-3 [NK]

Mouse, *Mtp-3*

Matsushima et al. (1984) used polyacrylamide gel electrophoresis to detect 15–20 protein components in the tears of various strains of mice. One of these, Mtp-3, is controlled by an X-linked gene, *Mtp-3*, which has 2 alleles, *Mtp-3ᵃ* and *Mtp-3ᵇ*. Of 20 strains surveyed, Matsushima and his colleagues found the former allele in only 1: New Zealand Black (NZB). Heterozygous females expressed mosaicism when tear samples from the right or left eye were examined separately.

Matsushima Y, Imai T, Ikemoto S: 1984. Evidence for X linkage of the mouse tear protein system-s (*Mtp-3*) with mosaic expression in heterozygous females. *Biochem Genet* 22: 577–585.

*31365 TEMPERATURE-SENSITIVE MUTATION (BA2R)

Mouse

The human X chromosome corrects the ts2 defect produced by ts2, a mouse 3T3 cell mutation which is temperature sensitive for cell and viral DNA synthesis (Jha et al. 1980). Nishimoto et al. (1982) isolated 219 ts mutants from the BHK cell line and classified them into 18 genetically defined complementation groups. Two of the mutations, BN462 (complementation group 3) and BN429 (group 6), are X-linked. The ts BN462 cell line has a ts defect in the progression of the G_1 to the S phase of the cell cycle (Sekiguchi et al. 1987). ts BN462 cells were transfected with the high-molecular-weight DNA from human KB cells, and a 56-kb DNA sequence, originating from the human X chromosome, was found to be conserved through 3 cycles of ts⁺ transformation.

Syrian hamster

Schwartz et al. (1979) described a recessive temperature-sensitive mutation of *BHK-21* hamster cells which was assigned to the X chromosome by complementation, karyotypic, and enzymatic analyses.

Jha KK, Siniscalco M, Ozer HL: 1980. Temperature-sensitive mutants of Balb/3T3 cells. III. Hybrids between *ts2* and other mouse mutant cells affected in DNA synthesis and correction of *ts2* defect by human X chromosomes. *Somatic Cell Genet* 6: 603–614.

Nishimoto T, Sekiguchi T, Kai R, Yamashita K, Takahashi T, Sekiguchi M: 1982. Large-scale selection and analysis of temperature-sensitive mutants for cell reproduction from BHK cells. *Somatic Cell Genet* 8: 811–824.

Schwartz HE, Moser GC, Holmes S, Meiss HK: 1979. Assignment of temperature-sensitive mutations of BHK cells to the X chromosome. *Somat Cell Genet* 5: 217–224.

Sekiguchi T, Yoshida MC, Sekiguchi M, Nishimoto T: 1987. Isolation of a human X chromosome-linked gene essential for progression from G1 to S phase of the cell cycle. *Exp Cell Res* 169: 395–407.

31384 TESTIS ASCORBIC ACID [NK]

Mouse, *Xta*

This was listed among the new gene symbols proposed in *Mouse News Letter* 71 as a "personal communication" to the editor. No further details are available.

Lyon JB: 1984. Personal communication to the editor. *Mouse News Lett* 71: 5 only.

*31386 TESTIS-DETERMINING FACTOR-LIKE SEQUENCES

Page et al. (1987) cloned a 230-kb segment of the human Y chromosome that contains some or all of the testis-determining factor gene (*TDF*). Certain DNA sequences within this region have been highly conserved during evolution; apparently homologous sequences occur on the Y chromosomes of several mammalian species examined. Very similar DNA sequences also occur on the X chromosome of humans and other mammals (listed in Table 3).

Page DC, Mosher R, Simpson EM, Fisher EMC, Mardon G, Pollack J, McGillivray B, de la

Chapelle A, Brown LG: 1987. The sex-determining region of the human Y chromosome encodes a finger protein. *Cell* 51: 1091–1104.

31387 TESTIS SIZE

Mouse

Hunt and Mittwoch (1987) investigated testis size in the BALB/c/Ola and CBA/Gr strains of mice. The former had larger testes from day 14 of embryonic development. Testis size was affected by the origin of the Y chromosome, the X chromosome, the autosomes (or pseudoautosomal region or both), and by maternal factors.

Hunt SE, Mittwoch U: 1987. Y-chromosomal and other factors on the development of testis size in mice. *Genet Res* 50: 205-211.

*31420 THYROXINE-BINDING GLOBULIN OF SERUM (TBG, SERUM)

Baboon

Lockwood et al. (1984) described X-linked polymorphism of TBG in the baboon. Two alleles, *TBG-C* ("common") and *TBG-S* ("slow"), are present in the proportions 0.72 and 0.28. A similar polymorphism exists in certain human populations. No such polymorphism was detected in 7 other primate species.

Lockwood DH, Coopenhaver DH, Ferrell RE, Daiger SP: 1984. X-linked, polymorphic genetic variation of thyroxin-binding globulin (TBG) in baboons and screening of additional primates. *Biochem Genet* 22: 81–88.

*31422 TISSUE INHIBITOR OF METALLOPROTEINASES [NK]

Mouse, *Timp*

The glycoprotein, tissue inhibitor of metalloproteinases (TIMP), is a physiological inhibitor of collagenase, stromelysin, and gelatinase, which are thought to be involved in extracellular matrix turnover (Jackson et al. 1987). The gene for human TIMP is X-linked (Huebner et al. 1986; Spurr et al. 1987), and Mullins et al. (1987) and Jackson et al. (1987) independently demonstrated that the gene is X-linked in the mouse. The locus is at the most distal part of the chromosome (Mullins et al. 1988).

Huebner K, Isobe M, Gasson JC, Golde DW, Croce CM: 1986. Localization of the gene encoding erythroid-potentiating activity to chromosome region Xp11.1→Xp11.4. *Am J Hum Genet* 38: 819–826.

Jackson IJ, LeCras TD, Docherty AJP: 1987. Assignment of the TIMP gene to the murine X-chromosome using an inter-species cross. *Nucleic Acis Res* 15: 4357 only.

Mullins LJ, Grant SG, Pazik J, Stephenson DA, Chapman VM: 1987. Assignment and linkage of four new genes to the mouse X chromosome. *Mouse News Lett* 77: 150–151.

Mullins LJ, Stephenson DA, Grant SG, Chapman VM: 1988. Detailed mapping of the proximal end of the mouse X chromosome. *Mouse News Lett* 80: 181–182.

Spurr NK, Goodfellow PN, Docherty AJP: 1987. Chromosomal assignment of the gene encoding the human tissue inhibitor of metallaproteinases to Xp11.1→11.4. *Ann Hum Genet* 51: 189–194.

*31433 TREMBLY [NK]

Mouse, *Ty*

The first mutants with this defect appeared among the descendants of a cross between 129/J and NZB/BINJ strains (Taylor et al. 1978). Affected males have tremors at about 2 weeks of age, seizures as the disease progresses, and do not survive beyond weaning. Histologically, their brains have none of the gross abnormalities that characterize jimpy animals (31208). Some presumptive heterozygous females exhibit mild tremors, poor coordination, and infertility. The locus is at the proximal end of the midregion of the chromosome between *Bpa* and *Bn*.

Taylor BA, Meier H, MacPike A, Williams M: 1978. Private communication. *Mouse News Lett* 59: 25 only.

*31434 TRF-ACCEPTOR SITE(S) ON B LYMPHOCYTES, EXPRESSION OF [NK]

Mouse

Tominaga et al. (1980) found that the responsiveness of B cells to T-cell replacing factor (TRF) is X-linked. Unlike most strains, which are high responders, the DBA/2Ha strain is a low responder, and subsequent studies indicated that DBA/2Ha B cells lack a TRF-acceptor site(s). The defective response in DBA/2Ha is a single recessive trait (Takatsu et al. 1981a). Takatsu and his colleagues have elaborated on this defective strain (Takatsu et al. 1981b; Takatsu and Hamaoka 1982). Sidman et al. (1986) found that the inability of B cells from these mice to respond to a similar family of lymphokines, B-cell maturation factors (BMFs), is due to the combined action of alleles at 2 autosomal loci: BMF responsiveness-1 (*Bmfr-1*) on chromosome 4 and BMF responsiveness-2 (*Bmfr-2*) on chromosome 9. They discuss several explanations for these discrepant results including the possibility that DBA/2Ha mice carry both autosomal and X-linked genes determining defective responses to B cells. In vivo, the TRF receptor deficiency causes only a very minor immunodeficiency (Baum et al. 1987): DBA/2Ha and DBA/2J mice have comparable numbers of phenotypically normal B lymphocytes; heterozygotes express both affected and normal B-cell populations; and DBA/2Ha mice respond well to a TI-2 antigent and a polyclonal activator.

Baum CM, Macke KA, Nahm MH: 1987. Study of DBA/2Ha immunodeficiency: X-chromosome mosaicism and in vivo immunoresponses. *Immunol Lett* 15: 179–185.

Sidman CL, Marshall JD, Beamer WG, Nadeau JH, Unanue ER: 1986. Two loci affecting B cell responses to B cell maturation factors. *J Exp Med* 163: 116–128.

Takatsu K, Hamaoka T: 1982. DBA/2Ha mice as a model of an X-linked immunodeficiency which is defective in the expression of TRF-acceptor site(s) on B lymphocytes. *Immunol Rev* 64: 25–55.

Takatsu K, Tominaga A, Hamaoka T: 1981a. X-linked recessive inheritance of a defective responsiveness to T-cell-replacing factor in DBA/2Ha mice. *Immunol Lett* 3: 137–143.

Takatsu K, Tanaka K, Tominaga A, Hamaoka T: 1981b. X-linked B cell defect in DBA/2Ha mice: low responder-status of B cells to T cell-replacing factor (TRF). In: Klinman NR, Mosier DR, Scher I, Vietta ES (eds), *B Lymphocytes in the Immune Response; Functional, Developmental, and Interactive Properties*. Amsterdam: Elsevier/North Holland, pp 331–338.

Tominaga A, Takatsu K, Hamaoka T: 1980. Antigen-induced T cell-replacing factor. II. X-linked gene control for the expression of TRF-acceptor site(s) on B lymphocytes and preparation of specific antiserum to that acceptor. *J Immunol* 124: 2423–2429.

*31453 VISUAL PIGMENTS, X-LINKED [30380/30390]

Visual pigments, light-absorbing molecules that mediate vision, consist of an apoprotein covalently linked to a small conjugated chromophore, 11-*cis*-retinal or, rarely, 11-*cis*-dehydroretinal. In vertebrates, they are located in the plasma and disk membranes of the photoreceptor outer segment. Each pigment is specified by its wavelength of maximal absorption; for the 3 pigments involved in human color vision, these are approximately 420 nm (blue-sensitive), 530 nm (green-sensitive), and 560 nm (red-sensitive). These pigments are located in the cone cells. Rhodopsin, the pigment involved in vision in dim light and absorbed maximally at 495 nm, is found in the rod cells. The visual pigments are structurally similar and are thought to have a common ancestor (Nathans et al. 1986a). In man, the gene determining the rhodopsin molecule on chromosome 7, that determining the blue pigment is on chromosome 3, and the remaining 2 are on the distal portion of the q arm of the X chromosome (Nathans et al. 1986b). The structure and sequencing of the 4 genes were described by Nathans and Hogness (1984), Nathans et al. (1986b), and Vollrath et al. (1988). The red- and green-pigment genes, which are virtually identical (98% sequence identity), are arranged tandemly in a head-to-tail fashion with the red at the 5′ end of the array. Only 1 red pigment gene is present, but multiple copies of the green pigment gene occur; these are assumed to have arisen by unequal crossing-over in the intergenic region. Such crossing-over could give rise to chromosomes possessing only the red-pigment gene. Alterations, involving unequal recombination between the very similar red- and green-pigment genes or gene conversion (or both), could explain the many variant forms of red–green color blindness in man. Old world monkeys appear to have the same green and red cone pigments as man, whereas New World monkeys have only a single, long-wavelength, visual pigment encoded on the X chromosome (Jacobs 1983). Evidence for the latter condition comes from studies, described below, on the squirrel monkey and the common marmoset. Nathans et al. (1986b) suggest that short-wavelength, long-wavelength (red and green), and rod pigments all diverged from a common ancestor about the same time, more than 500 million years ago. Separate red- and green-pigment genes were generated, following a duplication, after the split between New and Old World primates about 30–40 million years ago.

Mouse, red-sensitive visual pigment, *Rsvp*

Mullins et al. (1987) state that this locus is on the X chromosome closely linked to *G6pd* and *Cf-8*; no further details were given.

New World monkeys (Platyrrhina)

There are 2 platyrrhine families, Callirichidae and Cibidae. Color vision polymorphism is extensive in 1 cebid, the squirrel monkey (*Saimiri sciureus*) (Jacobs 1984), and similar variation may exist in other South American species of this family. About ⅓ of all squirrel monkeys are trichromatic, the rest dichromatic. There are 6 color-vision phenotypes, 3 dichromatic and 3 trichromatic, all of which have been observed (Bowmaker et al. 1987). Four types of cone pigments are found in the retina. The striking variations in color vision among these animals arise from individual variations in cone pigment complement (Mollon et al. 1984; Bowmaker et al. 1985). Although sample sizes are small, current evidence indicates that all male squirrel monkeys are dichromatic, whereas females are of both types, about ⅔ trichromatic (Jacobs and Neitz 1985a,b). These authors believe the most reasonable hypothesis compatible with current data is that only 1 X-linked gene, not as 2 as in man, controls the production of photopigments in this species. The locus in the squirrel monkey has 3 alleles, which occur with equal frequency. The inheritance of the cone pigment complements agrees with this hypothesis. Neitz et al. (1985) found preliminary evidence of a similar color vision polymorphism in a callithrichid monkey, and Travis et al. (1988) provided more extensive evidence in another callitrichid, the common marmoset, *Callithrix jacchus jacchus*. Although the details differ somewhat from those described in the squirrel monkey, the results of the microspectrophotometric measurements are generally consistent with the proposed model. How these various observations in the platyrrhines and the hypothesis derived from them fit in with findings derived from the molecular studies of Nathans et al. remains to be determined.

Bowmaker JK, Jacobs GH, Speigelhalter DJ, Mollon JD: 1985. Two types of trichromatic squirrel monkey share a pigment in the red-green spectral region. *Vision Res* 25: 1937–1944.

Bowmaker JK, Jacobs GH, Mollon JD: 1987. Polymorphism of photopigments in the squirrel monkey: a sixth phenotype. *Proc R Soc Lond [Biol]* 231: 383–390.

Jacobs GH: 1983. Variations in color vision among nonhuman primates. In: Mollon JD, Sharpe LT (eds), *Colour Vision: Physiology and Psychophysics*. London: Academic Press, pp 39–50.

Jacobs GH: 1984. Within-species variations in visual capacity among squirrel monkeys (*Saimiri sciureus*): color vision. *Vision Res* 24: 1267–1277.

Jacobs GH, Neitz J: 1985a. Color vision in squirrel monkeys: sex-related differences suggest the mode of inheritance. *Vision Res* 25: 141–143.

Jacobs GH, Neitz J: 1985b. Inheritance of color vision in a nonhuman primate. (Abstr) *Invest Ophthalmol Vis Sci (Suppl)* 26: 207 only.

Mollon JD, Bowmaker JK, Jacobs GH: 1984. Variations of colour vision in a New World primate can be explained by polymorphism of retinal pigments. *Proc R Soc Lond [Biol]* 222: 373–399.

Mullins LJ, Grant SG, Pazik J, Stephenson DA, Chapman VM: 1987. Assignment and linkage of four new genes to the mouse X chromosome. *Mouse News Lett* 77: 150–151.

Nathans J, Hogness DS: 1984. Isolation of nucleotide sequence of the gene encoding human rhodopsin. *Proc Natl Acad Sci USA* 81: 4851–4855.

Nathans J, Thomas D, Hogness DS: 1986a. Molecular genetics of human color vision: the genes encoding blue, green, and red pigments. *Science* 232: 193–202.

Nathans J, Piantanida TP, Eddy RL, Shows TB, Hogness DS: 1986b. Molecular genetics of inherited variation in human color vision. *Science* 232: 193–202.

Neitz J, Jacobs GH, Crognale M: 1985. Polymorphism of color vision in a callitrichid monkey. (Abstr) *Invest Ophthalmol Vis Sci (Suppl)* 26: 185 only.

Travis DS, Bowmaker JK, Mollon JD: 1988. Polymorphism of visual pigments in a callitrichid monkey. *Vision Res* 28: 481–490.

Vollrath D, Nathans J, Davis RW: 1988. Tandem array of human visual pigment genes at Xq28. *Science* 240: 1669–1672.

*31467 X-CHROMOSOME CONTROLLING ELEMENT (X-INACTIVATION CENTER)

Kangaroo (*Macropus robustus*)

Graves and Dawson (1988) demonstrated the existence of a single control locus on Xq near *HPRT* and *GLA* from which inactivation spreads outwards in both directions.

Mouse, *Xce*

While investigating variegation for autosomal coat color genes in female mice heterozygous for an X-autosome translocation, Cattanach and Isaacson (1965) observed that position effect variegation is under genetic control and that the responsible element is on the X chromosome (Cattanach 1966; Cattanach and Isaacson 1967). Subsequent studies on the modification of heterozygous phenotypes of *tabby* (30510) and the mottled allele *viable brindled* (30940) indicated that the controlling element is responsible for inactivation of the X chromosome, and that there are 2 alternate "states," or alleles, that permit complete (Xce^c) or incomplete (Xce^i) inactivation (Cattanach et al. 1969); this nomenclature was subsequently changed to Xce^a and Xce^b (Cattanach 1972). A third, more extreme, inactivation allele, Xce^c, was described by Johnston and Cattanach (1981). The Xce^c/Xce^a genotype causes nonrandom X^M-chromosome expression in the fetus and yolk sac mesoderm but not in the yolk sac endoderm at 13 ½ days post coitum (Bucher et al. 1985). The locus is in the midregion of the chromosome, very close to Ta (30510) (Cattanach et al. 1970; Cattanach and Papworth 1981; Cattanach 1983), and appears to influence all parts of it (Cattanach 1970). Cattanach and Williams (1972) observed X-linked modification of heterozygous phenotypes of Ta and Mo^{vbr} in the X chromosomes of several inbred strains; the modification appeared to result from nonrandom X chromosome inactivity; Cattanach (1974) elaborated upon and extended these observations. West and Chapman (1978) demonstrated similar Xce modification in 2 alleles at the $Pgk-1$ locus (31180); however, they could not determine whether the unbalanced mosaicism was the result of cell selection or nonrandom X chromosome inactivation; Johnston and Cattanach (1981) subsequently showed that the inactivation is nonrandom and occurs early, in the 7.5 day embryo. Ohno et al. (1973) and Drew et al. (1974) described modification of the Tfm locus (30494) and the Mo^{blo} allele (30940) by a controlling element that appears to be similar to or identical with Xce. The "inactivating center" postulated by Russell (1964: 732) and the controller gene (Cg) of Grahn et al. (1970) (later called "enhancer"; see Grahn 1972) are also probably identical with Xce.

Bucher T, Linke IM, Dunnwald M, West JD, Cattanach BM: 1985. Xce genotype has no impact on the effect of imprinting on X-chromosome expression in the mouse yolk sac endoderm. *Genet Res* 47: 43–48.

Cattanach BM: 1966. Genetic control over the spread of inactivation in a mouse X-autosome translocation. In: *Proceedings of the Symposium on the Mutation Process*. Praha: Akademia, pp 145–148.

Cattanach BM: 1970. Controlling elements in the mouse X-chromosome. III. Influence upon both parts of an X divided by rearrangement. *Genet Res* 16: 293–301.

Cattanach BM: 1972. Private communication. *Mouse News Lett* 47: 33 only.

Cattanach BM: 1974. Position effect variegation in the mouse. *Genet Res* 23: 291–306.

Cattanach BM: 1983. Private communication. *Mouse News Lett* 69: 24 only.

Cattanach BM, Isaacson JH: 1965. Genetic control over the inactivation of autosomal genes attached to the X-chromosome. *Z Vererb Lehre* 96: 313–323.

Cattanach BM, Isaacson JH: 1967. Controlling elements in the mouse. X-chromosome. *Genetics* 57: 331–346.

Cattanach BM, Papworth D: 1981. Controlling elements in the mouse. V. Linkage tests with X-linked genes. *Genet Res* 38: 57–70.

Cattanach BM, Williams CE: 1972. Evidence of non-random X chromosome activity in the mouse. *Genet Res* 19: 229–240.

Cattanach BM, Pollard CE, Perez JN: 1969. Controlling elements in the mouse X-chromosome. I. Interaction with the X-linked genes. *Genet Res* 14: 223–235.

Cattanach BM, Perez JN, Pollard CE: 1970. Controlling elements in the mouse X-chromosome. II. Location in the linkage map. *Genet Res* 15: 183–195.

Drew U, Blecher SR, Owen DA, Ohno S: 1974. Genetically directed preferential X-activation seen in mice. *Cell* 1: 3–8.

Grahn D: 1972. Private communication. *Mouse News Lett* 27: 21 only.

Grahn D, Lea RA, Hulesch J: 1970. Location of an X-inactivation controller gene on the normal X chromosome of the mouse. (Abstr) *Genetics* 64: s25 only.

Graves JAM, Dawson GW: 1988. The relationship between position and expression of genes on the kangaroo *X* chromosome suggests a tissue-specific spread of inactivation from a single control site. *Genet Res* 51: 103–109.

Johnston PG, Cattanach BM: 1981. Controlling elements in the mouse. IV. Evidence of non-random *X*-inactivation. *Genet Res* 37: 151–160.

Ohno S, Christian L, Attardi BJ, Kan J: 1973. Modification of expression of the *testicular feminization (Tfm)* gene on the mouse by a "controlling element" gene. *Nature [New Biol]* 245: 92–93.

Russell LB: 1964. Another look at the single-active-X hypothesis. *Trans NY Acad Sci, series II* 26: 727–736.

West JD, Chapman VM: 1978. Variation for *X* chromosome expression in mice detected by electrophoresis of phosphoglycerate kinase. *Genet Res* 32: 91–102.

*31440 XG BLOOD GROUP SYSTEM

In man this locus produces the Xg[a] antigen and controls 12E7 antigent expression (31347).

Gibbon

Gavin et al. (1964) tested several species of nonhuman primates and a number of mice and dogs for the Xg[a] antigen, and found it in gibbons only. Stone et al. (1979) could not detect Xg[a] in the rhesus monkey. The antigen in gibbons seems indistinguishable from that of humans, and its distribution in the small sample was compatible with X-linkage. Shaw et al. (1985) provided further evidence for X-linkage.

Gavin J, Noades J, Tippett P, Sanger R, Race RR: 1964. Blood group antigen Xg[a] in gibbons. *Nature* 204: 1322–1323.

Shaw MA, Tippett P, Delhanty JDA, Andrews M, Goodfellow P: 1985. Expression of Xg and the 12E7 antigen in primates. *J Immunogenet* 12: 115–118.

Stone WH, Sullivan PT, Blystad P: 1979. Immunogenetic studies of rhesus monkeys. 6. Absence of the human Xg[a] blood group on rhesus erythrocytes. *Anim Blood Groups Biochem Genet* 10: 57–59.

31491 XY SEX REVERSAL [?30610]

Horse

Kent et al. (1986, 1988) described this syndrome in which a phenotypic mare has the karyotype of a stallion; the disorder has been described in 3 breeds. Phenotypic expression, which ranges from a feminine mare with a reproductive tract within normal limits to a greatly masculinized mare, can be divided into 4 classes: I, nearly normal females, some of which are fertile; II, females with gonadal dysgenesis and normal Müllerian development; III, intersex mares with gonadal dysgenesis, abnormal Müllerian development, and an enlarged clitoris; IV, virilized intersex mares with high levels of testosterone. Of 38 affected mares described by Kent et al. (1988), 29 fell into classes II and III. In general, mares of classes I and II were H-Y antigen-negative, whereas those of classes III and IV were H-Y antigen-positive. Of 6 unrelated pedigrees described by Kent et al., only one is extensive. The authors conclude that the disorder is a sex-limited autosomal dominant or an X-linked recessive. The phenotypic variation is thought to be due to variable penetrance or to the effects of a modifier gene(s). The relation of this disorder(s) to similar ones in the human is not clear. XY gonadal dysgenesis (MIM 30610) is a human disorder in which phenotypic females with a 46,XY karyotype show bilateral streak gonads and sexual infantilism. Although it, too, exhibits genetic heterogeneity (Simpson et al. 1981), its phenotypic expression does not encompass the extremes in the horse.

Kent MG, Shoffner RN, Buoen L, Weber AF: 1986. XY sex-reversal syndrome in the domestic horse. *Cytogenet Cell Genet* 42: 8–18.

Kent MG, Shoffner RN, Hunter A, Elliston KO, Schroder W, Tolley E, Wachtel SS: 1988. XY sex reversal syndrome in the mare: clinical and behavioral studies, H-Y phenotype. *Hum Genet* 79: 321–328.

Simpson JL, Blagowidow N, Martin AO: 1981. XY gonadal dysgenesis: genetic heterogeneity based upon clinical observations, H-Y antigen status and segregation analysis. *Hum Genet* 58: 91–97.

*31493 X-Y UNIVALENCY, X-LINKED FACTORS CONTROLLING [NK]

Mouse, end-to-end association of the sex chromosomes (*Sxa*)

The frequency of X-Y univalency, which varies widely, appears to be strain specific (Gollapudi et al. 1981). In addition, an increase in the frequency of X-Y univalency appears to be a property of hybrids from crosses between laboratory strains (Rapp et al. 1977) and between laboratory strains and wild-caught subspecies (de Boer and Nijhoff 1981; Imai et al. 1981; Matsuda et al. 1982). These observations suggest that X-Y univalency is under genetic control. Matsuda et al. (1982) classified X-Y univalency (X//Y) into 3 types, 1 of which is found mainly in hybrids between phylogenetically distant subspecies, and has a high frequency (70–90%) of X-Y univalency. Matsuda et al. (1983) used crosses between BALB/c and *M. m. molossinus*, both of which have low X//Y (<30%), to analyze the genetics of this univalency. The frequency of X//Y was greater than 70% in the F_1 male hybrids, and both high and low types of progeny appeared in the backcross generation from BALB/c and high X//Y males; the frequency of low X//Y progeny decreased with each backcross generation. Low X//Y males produced only low X//Y offspring. These findings were explained on the basis of at least 1 gene (provisional symbol, *Sxa*) which is responsible for the end-to-end association of the X and Y, and is located on the common part of the X and Y chromosomes. It is postulated that the BALB/c animals are homozygous for an allele, *Sxa^a*, and that *M. m. molossinus* is homozygous for a second allele, *Sxa^b*. F_1 animals are *Sxa^a/Sxa^b*, and the segregation of low X//Y progeny from high X//Y sires is explained by crossing-over between the X and Y. The gradual decrease in low X//Y progeny during sequential backcrosses suggests the presence of an autosomal factor regulating crossing-over between the X and Y, and recombination of attachment sites. Biddle et al. (1985) demonstrated that the probability of univalency is regulated by 2 X-chromosome factors that differ between the C57BL/6J and DBA/2J strains. These factors appear to be separated by sufficient distance so that there are 2 recombinant classes of X-Y univalency at 20% and 60%. In this strain pair, 2 other systems appear to control univalency: 1 involves the Y chromosome, and the other an unknown number of autosomal factors. Biddle et al. suggest that these genetic systems influencing univalency may be concerned with the regulation or structure of the terminal attachment sites between the X and Y chromosomes.

Biddle, FG, MacDonald BG, Eales BA: 1985. Genetic control of sex-chromosomal univalency in the spermatocytes of C57BL/6J and DBA/2J mice. *Can J Genet Cytol* 27: 741–750.

de Boer P, Nijhoff JH: 1981. Incomplete sex chromosome pairing in oligospermic male hybrids of *Mus musculus* and *M. musculus molossinus* in relation to the source of the Y chromosome and the presence or absence of a reciprocal translocation. *J Reprod Fert* 62: 235–243.

Gollapudi BB, Kamra OP, Blecher SR: 1981. A search for a genetic basis for gonosomal univalency in mice. *Cytogenet Cell Genet* 29: 241–249.

Imai HT, Matsuda Y, Shiroishi T, Moriwaki K: 1981. High frequency of X-Y chromosome dissociation in primary spermatocytes of F_1 hybrids between Japanese wild mice (*Mus musculus molossinus*) and inbred laboratory mice. *Cytogenet Cell Genet* 29: 166–175.

Matsuda Y, Imai HT, Moriwaki K, Kondo K, Bonhomme F: 1982. X-Y chromosome dissociation in wild derived *Mus musculus* subspecies, laboratory mice, and their F_1 hybrids. *Cytogenet Cell Genet* 34: 241–252.

Matsuda Y, Imai HT, Moriwaki K, Kondo K: 1983. Modes of inheritance of X-Y dissociation in inter-subspecies hybrids between BALB/c mice and *Mus musculus molossinus*. *Cytogenet Cell Genet* 35: 309–315.

Rapp M, Therman E, Denniston C: 1977. Non-pairing of the X and Y chromosomes in the spermatocytes of BDF_1 mice. *Cytogenet Cell Genet* 19: 85–93.

*31498 Y CHROMOSOMAL SEQUENCES [NK]

Mouse

Nallaseth and Dewey (1986) described 3 cloned EcoRI genomic fragments whose sequences hybridize almost exclusively with male DNA. Male specificity is not absolute, and homologous sequences for 1 of the fragments, pBC 10-0.6, were found on the X chromosome.

Nonhuman primates

Koenig et al. (1984, 1985), Page et al. (1984), and Erickson (1987) have shown that certain human Y-chromosomal sequences are evolutionarily conserved among nonhuman primates; some of the sequences are on the X chromosome of these species.

Erickson RP: 1987. Evolution of four human Y chromosomal unique sequences. *J Mol Evol* 25: 300–307.

Koenig M, Camerino G, Heilig R, Mandel J-L: 1984. A DNA fragment from the human X chromosome short arm which detects a partially homologous sequence on the Y chromosomes short arm. *Nucleic Acids Res* 12: 4097–4109.

Koenig M, Moison JP, Heilig R, Mandel J-L: 1985. Homologies between X and Y chromosomes detected by DNA probes: localization and evolution. *Nucleic Acids Res* 13: 5485–5501.

Nallaseth FS, Dewey MJ: 1986. Moderately repeated mouse Y chromosal sequence families present distinct types of organization and evolutionary change. *Nucleic Acids Res* 14: 5295–5307.

Page DC, Harper ME, Love J, Botstein D: 1984. Occurrence of a transposition from the X-chromosome long arm to the Y-chromosome short arm during man evolution. *Nature* 311: 119–123.

*31499 YELLOW MOTTLING [NK]

Mouse, *Ym*

Hunsicker (1968) briefly described an X-linked lethal trait in which the mottling is yellowish and the vibrissae are not curly; the mutation is not allelic with Mo^{blo}.

Hunsicker PR: 1968. Private communication. *Mouse News Lett* 38: 31 only.

AUTHOR INDEX

SUBJECT INDEX

*COAGULATION FACTOR IX   30358
COLOR BLINDNESS   see VISUAL PIGMENTS, X-LINKED (31453)
COLOR VISION   see VISUAL PIGMENTS, X-LINKED (31453)
CONGENITIAL ECTODERMAL DEFECT (DOG)   see ECTODERMAL
   DYSPLASIA, ANHIDROTIC (30510)
*CONGENITIAL TREMOR TYPE AIII   30403
*CREAM   30413
CUTIS LAXA, X-LINKED   see MENKES SYNDROME (30940)
*CYSTINURIA   30432
*CYTOCHROME b-245, BETA POLYPEPTIDE   30433

DANCING PIG DISEASE   see CONGENITAL TREMOR TYPE AIII (30403)
DAPPLED (MOUSE)   see MENKES SYNDROME (30490)
*DIHYDROTESTOSTERONE RECEPTOR (TESTICULAR FEMINIZATION;
   ANDROGEN RECEPTOR DEFICIENCY; DHTR)   30494
DM-20 PROTEOLIPID PROTEIN   see PROTEOLIPID PROTEIN, MYELIN
   (31208)
DYSTROPHIN   see MUSCULAR DYSTROPHY, DUCHENNE TYPE (31010)

*ECTODERMAL DYSPLASIA, ANHIDROTIC   30510
*EYE–EAR REDUCTION   30536

FACTOR VIII DEFICIENCY   see COAGULATION FACTOR VIII (30356)
FACTOR IX DEFICIENCY   see COAGULATION FACTOR IX (30358)
*FIDGET, X-LINKED   30549
*FRAGILE SITE   30554
*FRAGILE SITE fra(X)(q1), FOLATE-SENSITIVE   30556
*FRAGILE SITE fra(X)(q2), FOLATE-SENSITIVE   30557
*FRAGILE SITE fra(X)(q3), FOLATE-SENSITIVE   30558
*FRAGILE SITE fra(X)(cen), FOLATE-SENSITIVE   30559
*FRAGILE SITE XC–D   30561

GINGER (CAT)   see ORANGE (31123)
*GLUCOSE-6-PHOSPHATE DEHYDROGENASE (G6PD, EC 1.1.1.49)   30590
GONAD-SPECIFIC RECEPTOR OF H–Y ANTIGEN   30615
GRANULOMATOUS DISEASE, CHRONIC   see CYTOCHROME b-245,
   BETA POLYPEPTIDE (30433)
*GREASY   30645
GYRO (MOUSE)   see HYPOPHOSPHATEMIA B, X-LINKED, TYPE II
   (30781)

*HARLEQUIN   30653
*HARVEY SARCOMA PROTOONCOGENE-2 PSEUDOGENE (HRAS2;

*LYMPHOCYTE ANTIGEN X-2   30907
*LYMPHOCYTE ANTIGEN X-3   30908
*LYMPHOCYTE-REGULATED, X-LINKED (XLR)   30909

MACULAR (MOUSE)   see MENKES SYNDROME (30940)
MARMALADE (CAT)   see ORANGE (31123)
*MENKES SYNDROME (KINKY-HAIR DISEASE)   30940
*MOLONEY MURINE LEUKEMIA VIRUS INTEGRATION SITE-14
   (M-MuLV INTEGRATION SITE-14)   30986
*MOLONEY MURINE LEUKEMIA VIRUS INTEGRATION SITE-15
   (M-MuLV INTEGRATION SITE-15)   30987
MONOCLONAL ANTIBODY 12E7: MIC2X   see SURFACE ANTIGEN MIC2
   (31347)
MOSIAC (MOUSE)   see MENKES SYNDROME (30940)
MOTTLED (HAMSTER, MOUSE)   see MENKES SYNDROME (30940)
*MUSCULAR DYSTROPHY, DUCHENNE TYPE (DMD)   31010
MYELIN DEFICIENCY (RAT)   see PROTEOLIPID PROTEIN, MYELIN
   (31208)
MYELIN SYNTHESIS DEFICIENCY (MOUSE)   see PROTEOLIPID
   PROTEIN, MYELIN (31208)
MYOCLONIA CONGENITA (PIG)   see CONGENITAL TREMOR TYPE AIII
   (30403)
MYOFIBRILLAR HYPOPLASIA (PIG)   see SPAYLEG (31336)
*NADH–COENZYME Q REDUCTASE DEFICIENCY (ELECTRON
   TRANSPORT CHAIN, DEFECT OF COMPLEX I OF)   31047
NEPHRITIS, HEREDITARY, X-LINKED   31048
NEUHERBERG (MOUSE)   see MENKES SYNDROME (30940)

OPSINS   see VISUAL PIGMENTS, X-LINKED (31453)
*ORANGE   31123
*ORINTHINE DECARBOXYLASE (OD;L-ORINTHINE CARBOXY-LYASE;
   EC 4.1.1.17)   31124
*ORINTHINE TRANSCARBAMYLASE (ORNITHINE
   CARBAMOYL-TRANSFERASE; OTC; EC 2.1.3.3)   31125
*OUABAIN RESISTANCE, X-LINKED (OUBR)   31135

*PARALYTIC TREMOR   31147
PATCHY COAT   31156
*PHOSPHOGLYCERATE KINASE 1 (PGK; EC 2.7.2.3)   31180
*PHOSPHORYLASE KINASE (DEPHOSPHOPHOSPHORYLASE KINASE;
   ATP: PHOSPHORYLASE-6-PHOSPHOTRANSFERASE; PHK; EC
   2.7.1.38)   31187
POLYDACTYLY   31201

Printed in the United States
By Bookmasters